CAVE OF BONES

Also by Lee Berger and John Hawks
Almost Human

Also by Lee Berger
The Skull in the Rock (with Marc Aronson)
In the Footsteps of Eve (with Brett Hilton-Barber)

CAVE OF BONES

A TRUE STORY OF DISCOVERY, ADVENTURE, AND HUMAN ORIGINS

LEE BERGER
AND JOHN HAWKS

Washington, D.C.

Published by National Geographic Partners, LLC
1145 17th Street NW Washington, DC 20036

Copyright © 2023 Lee Berger. All rights reserved. Reproduction of the whole or any part of the contents without written permission from the publisher is prohibited.

NATIONAL GEOGRAPHIC and Yellow Border Design are trademarks of the National Geographic Society, used under license.

Library of Congress Cataloging-in-Publication Data

Names: Berger, Lee R., author. | Hawks, John (John David) author.
Title: Cave of bones : a true story of discovery, adventure, and human origins / Lee Berger, John Hawks.
Description: Washington, D.C. : National Geographic, 2023. | Includes bibliographical references and index. | Summary: "This thrilling book takes the reader into South African caves to discover fossil remains that reframe the human family tree"-- Provided by publisher.
Identifiers: LCCN 2023017261 (print) | LCCN 2023017262 (ebook) | ISBN 9781426223884 (hardcover) | ISBN 9781426223945 (ebook)
Subjects: LCSH: Homo naledi. | Human beings--Origin. | Human remains (Archaeology)--South Africa--Witwatersrand Region.
Classification: LCC GN284.5 .B473 2023 (print) | LCC GN284.5 (ebook) | DDC 569.9--dc23/eng/20230508
LC record available at https://lccn.loc.gov/2023017261
LC ebook record available at https://lccn.loc.gov/2023017262

Since 1888, the National Geographic Society has funded more than 14,000 research, conservation, education, and storytelling projects around the world. National Geographic Partners distributes a portion of the funds it receives from your purchase to National Geographic Society to support programs including the conservation of animals and their habitats.

Get closer to National Geographic Explorers and photographers, and connect with our global community. Join us today at nationalgeographic.org/joinus

For rights or permissions inquiries, please contact National Geographic Books Subsidiary Rights: bookrights@natgeo.com

Interior design: Lisa Monias

Hardcover ISBN 978-1-4262-2388-4
Export edition ISBN 978-1-4262-2414-0

Printed in the United States of America
23/WOR/1

To the spirit of exploration and explorers everywhere.
Never stop exploring.

CONTENTS

Prologue 9

PART I: THE *NALEDI* JOURNEY 15
1. The Cradle of Humankind 17
2. Our Family Tree 25
3. Finding *Homo naledi* 45
4. The World Meets *naledi* 57
5. Chamber of Many Stars 63

PART II: SO MANY BONES 69
6. Inside the Lesedi Chamber 71
7. Cave Dwellers 77
8. Another Body 81
9. Hints of Burial 89
10. A Turning Point 101

PART III: JOURNEY INTO DARKNESS 113
11. Training 115
12. Approaching the Chute 123
13. Into the Chute 133
14. The Descent 139
15. Exploring the Hill Antechamber 147
16. The Markings 151
17. More Markings 159
18. Struggling Out 171

PART IV: MEANING 179
19. Markings and Meaning 181
20. Burnt Bone 187
21. Traces of Culture 191
22. The Search for Meaning 199

Epilogue 207

Acknowledgments 213
Appendix A: Known Humans Who Have Entered the Dinaledi Chamber .. 217
Appendix B: Time Line of *Homo naledi* Discoveries 218
Bibliography 221
Illustrations Credits 227
Index 229

PROLOGUE

There is always a moment of doubt before doing something dangerous—and I had plenty of doubt as my feet slid into the narrow abyss we call the Chute. I was face up against solid rock, with my blue cotton jumpsuit snagging on the crags and points in the stone and my legs dangling under my body, thighs barely fitting inside the fissure. My helmet light cast eerie shadows around me—it could penetrate only five meters ahead into the darkness.

For nine years, from 2013 to 2022, I had watched others slip through this same fissure, down the Chute on the way into a space we had dubbed the Dinaledi Chamber—the cavernous location of a rich cache of fossil bones that had become the research focus of not just our expedition team but also scientists from around the world. I had always viewed this cave through a computer screen safely situated in the Command Center near the entrance to the caves. The Command Center's cavern is large enough to house plastic chairs and tables, and from its relative comforts I could view activity going on deeper underground, the action reaching my screen via cables we'd strung through the cave system over the years.

But this time, I was making the journey into the Dinaledi Chamber myself. In recent months, my team had uncovered an astounding trail of clues that had the potential to revolutionize our research into human origins. We seemed to be on the cusp of new understanding about ancient human relatives and, in turn, about who the human species is today. So despite the danger, I was leaving the Command Center and braving the Chute to reach the remarkable underground space that had already, years before, stunned the world with its trove of fossil evidence—the richest site of prehuman remains ever discovered. With its thousands of bones, this cave system has rewritten the story of human ancestry. It has also changed the course of my life.

Over two expeditions totaling seven weeks across 2013 and 2014, my team members had recovered more than 1,200 fossils, primarily of bones and teeth from an area in Dinaledi no bigger than one square meter. I often say that before this discovery, the number of paleoanthropologists—or people who study hominins—around the world exceeded the number of bones we had to study, but our finds changed that. After we described these findings in more than a dozen scientific papers, our team of experts in ancient hominin anatomy reported that these fossils were unlike anything paleoanthropologists had ever seen at any other site. The fossils represented a new species, a new ancient human relative that we named *Homo naledi*—*Homo*, because our analysis determined that it belonged in the genus shared by other species closely related to humans; and *naledi*, meaning "star" in Sesotho, a common language of this region of South Africa.

But during the *Homo naledi* breakthrough and beyond, fewer than 50 of my coworkers had shimmied the 12 meters down the Chute to Dinaledi and, despite leading this research for nearly a decade, I had pictured the space only in my imagination, filling in details by watching others on the computer screen, listening to them describe the space, reviewing maps, and marveling at the fossils they had excavated and laboriously carried up to the Command Center. I had told thousands of people about the

PROLOGUE

perils of this space over the years. But until now, I had never been inside the Dinaledi Chamber myself.

With my lower body already inside the tunnel, I took a deep breath, the last I would be able to take for some time, and envisioned the narrow confines I was entering. The slimmest part of the Chute was just 19 centimeters across—about seven and a half inches, or as wide as the short side of a shoebox. Would I fit? And if I made it down to Dinaledi, could I get back out? I was turning 57 in a few months, and though I was strong, I was never a person anyone would call slender. I'd lost a lot of weight for this attempt, but was it enough?

I pushed against the ancient gray rock and felt my pelvis enter the Chute. Damn, this is tight, I thought. As I wriggled farther, my feet sought the rocky knobs I knew to be at the top of the shaft. I found only one. I braced my foot against it, then let the other leg dangle. I held my breath, steeled myself, then pushed off with my hands, letting gravity pull my hips down through the gap. A sharp rock scraped my belly. I dangled half in and half out of the chimneylike opening. This was just the beginning.

I looked up at Maropeng Ramalepa, a member of my exploration team and my guide for this first half of the descent. He crouched at the opening to the tunnel in the spot we had come to call the Chute Troll position. He had made this journey dozens of times. He offered a broad smile, and his eyes sparkled in the light of my headlamp. "You got this, Prof!" he said.

I answered with a grunt. My breath was already steamy in the cool cave air. I gingerly moved my boots about for more toeholds, lowering myself farther until my chest reached the same squeeze that my hips had just cleared. Rock pressed hard against my spine and sternum.

I sucked in my gut and exhaled to make my chest smaller, then shoved myself downward again to force my chest through the impossibly small gap. I felt a sharp pain as a rock pierced my upper back, but then I made it. My whole body was in the Chute.

My arms stretched above my head, and my feet scrambled for purchase below me. I wanted to look down to see where I was going, but that just resulted in scraping my helmet against the rock. Looking up, I could see the narrow gap I had just struggled through. I would eventually have to force my way back up and out of there. Doubt crept in, but my rational mind took charge: I had questions to answer, perhaps discoveries to make. It was finally time for me to witness in person the most important site of my career.

I tried to take one more deep breath, but the walls of rock restricted my inhalation. Nevertheless, I was determined. My fingers searched for the next handhold as I inched myself lower through the Chute. Little did I know I was headed for some of the most wondrous and terrifying moments of my life.

PART I

THE *NALEDI* JOURNEY

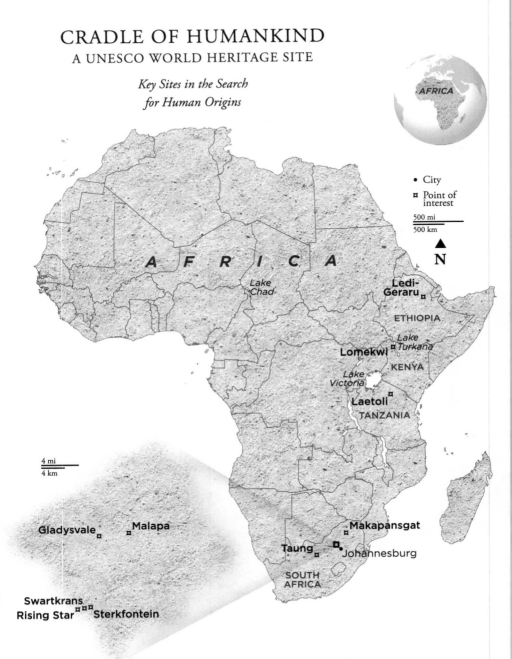

The Rising Star cave system is one of many early hominin sites in South Africa's Cradle of Humankind.

| 1 |

THE CRADLE OF HUMANKIND

Intrepid cavers have known about the rocky passageways of Rising Star, the primary cave system where my team studies hominins, for more than 40 years. This system, including the Chute and the Dinaledi Chamber, is one of many such cave networks in the region of South Africa designated the Cradle of Humankind UNESCO World Heritage site. It's a landscape of rolling grasslands, small farm holdings, and game reserves cut with streams, rivers, and the occasional cluster of trees. South Africans call this region the Witwatersrand, or "ridge with white waters," for the many waterfalls that cascade over the low-lying hills. An hour's drive from Johannesburg, the Witwatersrand sits within the Highveld, a greater region of central southern Africa that rises to an altitude of 1,525 meters (5,000 feet) above sea level, a plateau above the coastal regions and deserts to the north, southeast, and west.

Bedrock pokes through the surface here and there on the Highveld, leaving little space for topsoil to accumulate and rendering the plateau's terrain largely useless for crop farming. The region's rocky underlay is

Rising Star team member Mathabela Tsikoane sits among typical flowstone and stalactite formations in the cave system.

composed of an extremely hard dolomitic limestone, a gray sedimentary rock rich in both calcium carbonate and magnesium, that was formed by the slow accumulation of lime and sand on shallow warm-water seas between two and three billion years ago, long before complex life evolved. The rock in these caves looks like a layer cake, with pure dolomite layers separated by thin layers of chert, an ultrahard, brittle sedimentary rock made mostly of silica. Chert's darker, glassy appearance contrasts starkly with the dull matte texture of the thicker layers of dolomite.

Calcium carbonate dissolves out of the dolomite over long periods of time, and dripping water forms it into pure white layers we call flowstone, or into stalactites and stalagmites: respectively, those pillars that hang from the ceiling or rise from the floor. In the early 20th century, miners went to enormous lengths to harvest calcium carbonate from these caves, which was then heated in great kilns to create the agricultural chemical we colloquially call lime. Occasionally, in pockets and in the larger spaces of these caves, chunks of material coalesce, cemented together by the dripping calcium carbonate. These deposits, called breccia, sometimes contain fossils of ancient life-forms that lived in this region and these caves. The breccia and flowstone layers here can be anywhere from a few thousand to more than three million years old.

Greater rewards than lime lie hidden beneath this unassuming landscape. Investigate any of the sparse tree clusters, and you are bound to find a depression created by an underground collapse. In extreme cases, sinkholes up to 30 meters deep hint at massive underground caverns. Sometimes horizontal caves, from large and imposing caverns to rocky nooks, have formed. Intricate networks of passages, some explored but many inaccessible, thread beneath this high plateau. Keen-eyed explorers can find an entrance, often just a hole, into these vast underground labyrinths—an entrance like the one I ventured into at Rising Star.

Tremendous events like earthquakes and meteorite impacts have shaped this underground landscape, creating paths for water to eat away at the dolomite. Slower, gentler processes, such as plant roots invading tiny cracks in the rock, have created fissures as well. Underground rivers erode and shape rock over millennia; sudden falls or collapses form larger spaces. Understanding all these geological forces helps us theorize what this underground world might have been like hundreds of thousands of years ago—including how it might have accommodated human ancestors.

The Rising Star cave system comprises nearly four kilometers (2.5 miles) of interlaced passages, descending in some places as deep as the water table more than 40 meters belowground. Of these four kilometers, a 250-by-150-meter portion is accessible, and cavers and explorers have mapped this section extensively. An experienced caver can explore this complex system by traveling along latticelike shelves of limestone and chert and squeezing through openings in the stone. Occasionally, you might find a chamber in which you can sit up or even stand, but most of the open spaces are relatively small, often less than a meter wide. Some of the larger chambers in Rising Star reveal extraordinary cave formations, stalactites glistening like crystal chandeliers from the ceiling, and stalagmites rising like pillars from the floor.

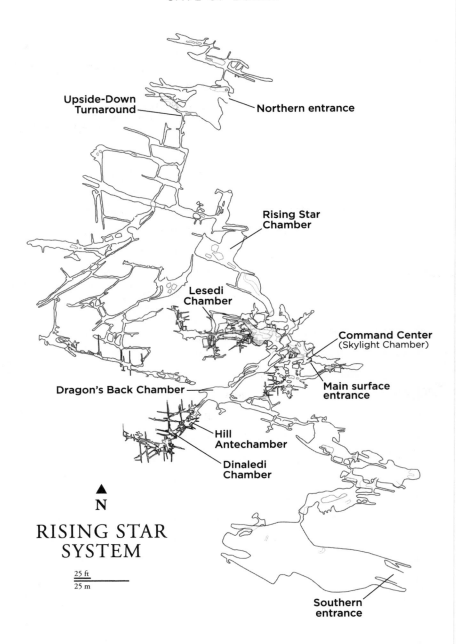

An extensive network of openings and passages make up the Rising Star cave system.

THE CRADLE OF HUMANKIND

Caves have long been part of the deep human journey, whether as places of refuge or places of death. We've memorialized that part of our ancient history in the term "caveman," but the story itself is more complicated.

The remains of some of our earliest ancestors, the australopithecines—small-brained hominins that lived in this region more than two and a half million years ago—have been found in caves. We don't know the exact interactions these earliest ancestors had with the subterranean world. Some of their bones bear the marks of great predators, such as saber-toothed cats and leopards, who left their remains in the caves' dark recesses. Later hominins, such as the species *Homo erectus,* which lived more than a million years ago, had a more congenial relationship with underground spaces. Found in Asia, Europe, and Africa, they left stone tools and butchered bones near the entrances of caves, sometimes together with evidence of fire.

Caves can offer refuge from predators, a place of safety, a cool place in a hot climate. With the acquisition of fire, a cave can be warm in winter, offering shelter from wind, rain, and lightning. The prehistoric use of caves, together with the fact that they tend to protect bones and artifacts for long periods of time, is why many people think of our ancient ancestors as "cavemen." But with a few exceptions in Europe among Neanderthals and early *Homo sapiens,* most ancient hominins' use of caves is ephemeral, confined to only the caves' shallowest parts. Ancient people hovered near cave entrances in rock shelters and overhangs and often avoided the deeper, darker spaces where light does not penetrate.

As a paleoanthropologist working primarily in South Africa, I have spent a lot of time exploring caves. My experience leads me to divide these subterranean spaces into three categories: the living, the dead, and the touched.

The living zones are the first parts of the cave that you enter from outside, and they extend to the spaces just beyond, where larger living organisms venture. These spaces teem with the smells, sounds, and sights of an active ecosystem: the buttered-popcorn smell of leopard urine, a porcupine's musky scent, blooming mold from fallen and rotting vegetation. You can hear Cape robins singing in the alcoves and barn owls ruffling their feathers on a rocky perch. But once you venture deeper into the system, the light fades away and the world becomes dim and dark—you need a headlamp from here.

The transition between the living zone and the ensuing dead zones can be 50 meters or more in distance. Here, bats might flit past your face, and your light might illuminate the reflective eyes of a cave spider, but the world generally narrows down to only what's captured by your circular headlamp beam. Everything outside that cone of light is invisible. The confines deaden any stray sounds, too—even the voices of nearby companions. The only sounds you hear are those you create—breathing in and out, gloves brushing against damp rock, boots scraping along a ledge. You feel as if you are always descending.

These spaces, like the passages and chambers of Rising Star, aren't truly dead—they're alive in their own way, but it is an inorganic life. The caves seem to breathe, literally, as air moves through each small crevice—pressure differentials between larger unseen spaces push air around, in and out of unseen surface openings. Although spaces in the dead zones are often relatively dry, they always feel humid, particularly chambers like Dinaledi where the stalactites and stalagmites glisten like diamonds when your light flashes over them—water droplets trickling down toward the water table below. With every drip, the water leaves tiny deposits of lime. Here, you are truly watching creation happen. Each white spindle or flat flowstone slab takes thousands of years to form.

As you clamber through passageways from chamber to chamber, your sense of touch and your awareness of pressure become incredibly important. You test every handhold. Can it bear your weight? Is there enough traction for your gloved hand to cling to this point if you begin to fall? You test and retest every foot placement. There can be few leaps of faith; an injury here can go from inconvenient to fatal. These journeys happen in slow motion. Even in Rising Star, some of these spaces have been visited by humans fewer times than the moon has.

The "touched" spaces of a cave are in many ways the most significant for paleoanthropologists. These spaces hold evidence that humans or hominins have been there before, or they are tragic places where an animal has ventured too far in and died of starvation or from a fall, almost always alone. I feel a strange array of emotions when I crawl into a space and find something like a deceased baboon or a dead honey badger. I think about how this poor animal came to be so far back in a dark, mysterious space. What was it searching for? Was it lost or confused? Now its body lies flat and mummifying, little left but dried skin, fur, and bones. The space has been touched by death. It is forever changed.

This is the curse of studying ancient species. We explore the places where things have died. This is often how fossils form. They begin in tragedy, but eventually they achieve an immortality that the billions who live and die on the surface will likely never attain.

| 2 |

OUR FAMILY TREE

It's time for a brief discussion of the study of human evolution and where it stands today. When I give talks, everyone wants the snapshot version, the easy answer to the implicit question: How is *Homo naledi* related to me? The fact is, there's no simple answer. I'm going to turn the mic over to John Hawks at this point, because he's got a better way of answering that question than I do.

Lee and the rest of our team are at the center of a fast-moving science, often making discoveries that surprise us. Even things we learned about human origins years ago have changed since we learned them.

When I was a kid in the 1970s and early '80s, I loved to read books about science. I must have checked out every book the library had on the subject. In the small town where I grew up, the library had plenty of kids' books on dinosaurs and sea life, but to read about human evolution, I

For decades, human evolution was portrayed as a linear "march of progress"

had to go to the grown-up section. One of the books I remember, titled *Early Man,* had pages that folded out to reveal a picture of a detailed painting depicting every species of ancient human relative known at the time: 15 figures walking in a line, each one step behind the other.

Homo sapiens was at the head of the line, looking like he just stepped out of a physical exam. He was followed by a Cro-Magnon man, who held a spear, then Neanderthal, then *Homo erectus, Australopithecus africanus,* and so on. At the far left of the parade—the very back of the line—was a small ape, its arms teetering outward as if it were walking a tightrope. Every one of them was a male figure, and from right to left, front to back, present to past, they gradually became shorter, hairier, and more hunched over. When you folded the pages back into the book, you saw only six figures: the most primitive and the most advanced, seemingly all in a row.

This picture, often called "The March of Progress," is one of the most famous images in the history of science. You've seen those figures on T-shirts and billboards, as marketing for companies or museums. Often somebody will add a figure at the front of the line to show the next stage of human progress—a person sitting hunched at a computer, or wearing a space suit, or sporting a beer belly. It's iconic, it's everywhere—and it's wrong.

OUR FAMILY TREE

from ape to modern human. Current science shows it's not that simple.

We did not evolve as a single line of progress. Our fossil relatives form a large branching tree, not a parade. Even in the 1960s, when *Early Man* was published, paleoanthropologists knew that our ancestors and relatives were a diverse lot. Sure, scientists faced many challenges in working out exactly how the fossils were related to each other, and a lot of uncertainty remains today. But nobody thought that the fossils could all be arranged into a single line of ancestors and descendants. The scientists knew that some fossils had features similar to those of modern humans, often in important ways, but they represented species that had gone off on their own paths. Each of these relatives is one branch of the great tree that connects us with them and with other living and ancient primates.

Fossils represent one kind of evidence about the shape of this evolutionary tree. More and more, though, we are also relying on another kind of evidence: the DNA record from both ancient and living species.

Before I started on the Rising Star journey with Lee, most of my research had been in genetics. I worked to understand how humans around the world today are related to each other and how ancient groups might fit into the big picture. For those questions (and others), DNA is not better or worse than fossils; each kind of evidence gives us different information. DNA can be especially useful, though, for working out how different branches first diverged from each other, tracing back to the time that they last shared a common ancestor. In very much the same way that you can learn about your family tree by sending a saliva sample for analysis, we can do similar comparisons among the genomes of other living primates.

The closest living relatives of humans are chimpanzees and bonobos, two closely related species of African apes. These two came from one ancestral stem around two million years ago. That stem and our own branch of the primate tree are sisters. We know that a last common ancestor of these three species existed, and DNA from living humans, chimpanzees, and bonobos tells us that this common ancestor lived sometime between six and eight million years ago. We have not yet found fossils of this common ancestor, but the other ape fossils from this time and earlier show us a great diversity of anatomies, suggesting many different ways of life. The hominin tree, including people today and all our fossil hominin relatives, started from this time forward.

For more than a decade now, although I continue to do work on genetics, I've spent most of my time working with fossils, and a lot of it in the fossil vault at the University of the Witwatersrand in Johannesburg. It's a special place, with thousands of pieces of ancient hominin bone from more than a dozen sites in South Africa. One of the oldest and most significant hominin fossils rests in a glass museum case specially made for it. When the anatomist Raymond Dart discovered this fossil skull in 1924, he named it *Australopithecus africanus*. Nowadays we often call it the Taung Child, for the site where it was found. Later discoveries from other South African sites such as Sterkfontein and Makapansgat revealed more

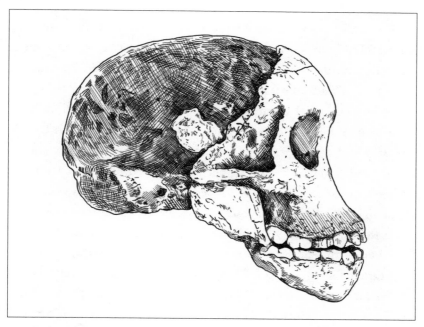

Found in South Africa in 1924, the skull of the Taung Child (Australopithecus africanus) *hinted at the region's rich early hominin history.*

and more skeletal remains of *africanus*. Today we believe that this species lived some 3 million to 2.1 million years ago.

While each find represented only a small part of a body, as paleoanthropologists combined them, the picture became clear. *Australopithecus* was a biped, able to walk upright like humans. The shape of its spine, legs, feet, and pelvis all helped it walk and run bipedally and hindered it from moving on all fours. But its brain was much smaller than a human's: around 400 to 500 cubic centimeters, compared with our average brain size of 1,400 cubic centimeters. These beings had much bigger, thicker molar and premolar teeth than we do, and researchers are still trying to work out what their diet was. Their faces were different from ape faces in one important way: *Australopithecus* had much smaller canine teeth than apes. But otherwise, the *Australopithecus* face might remind you more of a chimpanzee's or gorilla's than a modern human being's.

Many of the celebrated fossils representing the hominin record are in the genus *Australopithecus*. In the 1970s, teams working in Ethiopia and in Tanzania unearthed an early form of this genus, *Australopithecus afarensis*. This species includes the famous partial skeleton popularly known as Lucy, found by Donald Johanson and Tom Gray in 1974. *Afarensis* probably made most of the fossil footprints uncovered by Mary Leakey at Laetoli, Tanzania, in 1976. Most *afarensis* fossils have been dated to between 3.6 million and 3 million years ago, but the oldest go back to as far as 3.9 million years ago. These discoveries were brand-new when I was a kid.

But when I was in graduate school, scientists pushed the record back much further. By the 1990s, Meave Leakey was leading a team of fossil hunters near the southwest corner of Lake Turkana, Kenya, and found even earlier fossils, with a more apelike jaw shape, which they named *Australopithecus anamensis*. Other scientists found more fossils of this species in Ethiopia, including a skull. The fossils of *anamensis* are between 4.2 million and 3.8 million years old.

Each of these discoveries—*africanus, afarensis, anamensis*—represents earlier and earlier species, but they all were alike in many ways: all bipeds, all averaging smaller in weight and shorter in stature than humans, all with smallish brains. There were some variations on the theme: longer arms and bigger molars in *africanus,* bigger jaw muscles and body size in some *afarensis* individuals, and a more apelike jaw and face in *anamensis.* Clearly something clicked between these early relatives and their environments to enable them to succeed for some two million years. But they were not alone.

Further fossil finds reveal to us that through much of this time, other hominin species existed in Africa—species we know much less about: *Kenyanthropus platyops* from the area around Lake Turkana, *Australopithecus deyiremeda* from Ethiopia at the same time as *afarensis, Australopithecus garhi* from a bit later, and another hominin that we know so little about that scientists haven't even named it. This last one seems to have had an opposable big toe—a trait that suggests some different form

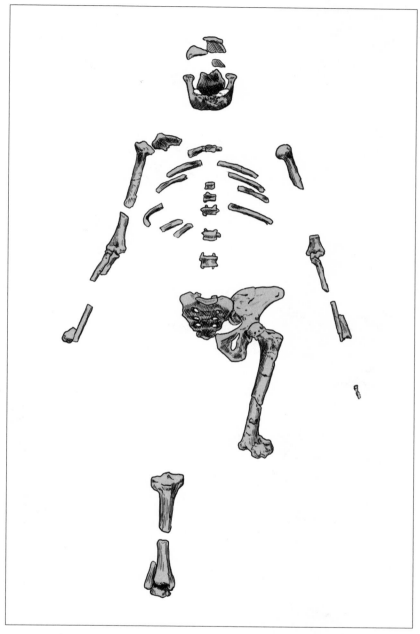

The fossils of so-called Lucy (Australopithecus afarensis) *represented nearly half a skeleton, found in Ethiopia in 1974.*

Today's panorama of hominin fossils represents the wide array of human ancestry, more a branching network than a single evolutionary line.

Kenyanthropus platyops

Australopithecus afarensis

Ardipithecus ramidus

Australopithecus africanus

Australopithecus anamensis

Paranthropus aethiopicus

4 million　　　3

OUR FAMILY TREE

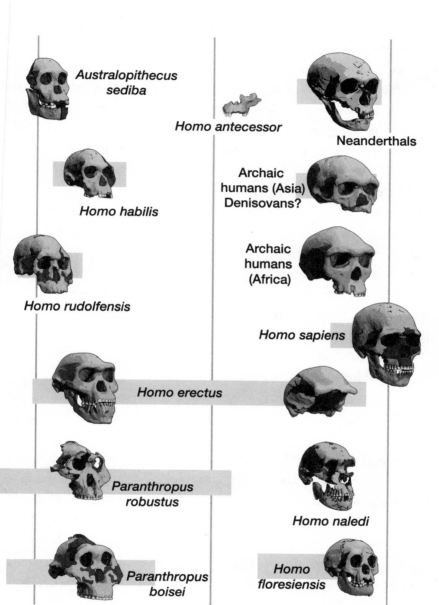

33

of walking and climbing. Each of these species is represented by bones and teeth that don't fit very well with the better-known species, such as *afarensis* and *africanus*.

By this point, you're probably wondering how we can be so sure that all these species were really different from each other. Isn't it possible that some of them were just slightly different versions of the same thing?

The answer may surprise you. Paleoanthropologists argue vociferously about almost every species—its identity, its boundaries, and whether it actually differs from others. When the fossil evidence is slim, it's hard to answer these questions. For example, *Kenyanthropus platyops* is known from a few dozen teeth, jaw, and skull fragments, plus one very distorted skull, all around 3.5 million to 3.3 million years old. The fossils show some notable differences from *afarensis,* which is the best known species from the same time. But some paleoanthropologists have argued pretty strongly that those differences don't matter, and that the *Kenyanthropus* fossils represent a regional population of *afarensis*.

And I could point to the same kind of disagreement for almost every species that has been discovered.

By studying distinctions among the fossils, we are working to understand how diversity evolved. The depth of time from the common ancestors of different groups, the store of genetic variation they may have had, and their chance of getting new genes by interbreeding with other populations—all these things matter. I imagine that if we could go back in time and watch these hominins, we would find them preferring slightly different habitats, eating a different mix of foods, and behaving in some noticeably different ways. They were all close relatives, but they evolved and diversified over the course of two million years or more.

We can look at living primates to get an idea of how such closely related species can differ from each other. Living chimpanzees and bonobos share a common ancestor back two million years, and today they have very different social systems and some noticeable differences in anatomy. Six species of baboons live across the African continent today, all evolved from a common ancestor around two million years ago. Those six baboon species share a great deal in the way of behavior, diet, and body shape, but they differ in social behaviors, fur patterns and coloration, and body size, in addition to having many other slight anatomical distinctions. For these living species—and for fossil hominins that existed within the last half million years—DNA now tells us a great deal about their interactions and histories, including occasional hybridization. But for more ancient fossils, like *Australopithecus,* we have only the shapes of their bones.

The earliest known *Australopithecus* fossils come from at least two million years after the hominin tree arose, and maybe as many as four million. This is an immense span of time with very few fossils. Field researchers have found four other fossil species from this lengthy period that may also belong to our hominin tree, each quite different from *Australopithecus*. All four species have relatively small canine teeth, compared to those of living great apes, and their fossils hint that they may have been better than living apes at bipedal standing or walking.

The one we know the most about is *Ardipithecus ramidus*. Most of its fossils are about 4.4 million years old—just 200,000 years older than the earliest *Australopithecus*—and were discovered in the same part of Ethiopia where *anamensis* and *afarensis* fossils have been found. But *Ardipithecus ramidus* was markedly different from the later *Australopithecus* species. It had an opposable big toe, long arms, elongated fingers, and a short thumb. Its body might remind you of a bonobo's. Its molar and premolar teeth were smaller, with thinner enamel, than those of *Australopithecus*. Since so much of its anatomy is like that of living great apes, some scientists wonder whether *Ardipithecus ramidus* might actually have been part of the chimpanzee branch, or another extinct branch of the apes.

Still, the genus *Ardipithecus* shares quite a few features with *Australopithecus*. The base of its skull and shape of its pelvis suggest that it held its spine upright a lot of the time, and its canine teeth were a bit smaller than those of most modern apes. In short, *Ardipithecus ramidus* seems like an ape that was testing the bipedal lifestyle—what many of the first members of our branch must have done.

Compared with *ramidus,* earlier fossil species are a lot sketchier. *Ardipithecus kadabba,* a million years earlier (hence 5.5 million years old), is represented by only a few fossils that suggest its teeth differed in a few details from *ramidus* teeth. *Orrorin tugenensis,* around six million years old, from western Kenya, is represented by fragments of three thighbones, shaped a little differently from those of the great apes, suggesting more frequent bipedal postures. Earliest of all is *Sahelanthropus tchadensis,* from seven million years ago in the area near present-day Lake Chad in north-central Africa. The fossils include a skull and a portion of its thighbone, and scientists disagree about how this species moved and whether it might represent another branch of apes entirely.

Lee and I both agree that these early fossils before *anamensis* are not enough evidence to understand how our branch of the primates, the hominin branch, got its start. Maybe all of them belong to extinct groups of apes that coexisted with our early ancestors. Maybe becoming a biped was not a single event. Our way of movement emerged from millions of years of experimenting by creatures that relied on trees for shelter and ate forest and woodland foods. We still don't know exactly why *Australopithecus* and its bipedal cousins lasted, but clearly it was not a linear march of progress! This earliest phase of our evolution was unquestionably important, setting our lineage apart from the great apes', but it was complicated, too. We still have much more to learn.

The second half of our evolutionary story—the last three million years—was also a drama with many players coming on and off the stage. The events in this book, including the discoveries made about *naledi,* come near the end of the story, within the last few hundred thousand years. But their roots lie much deeper. Our genus, the one we know as *Homo*, arose sometime between three million and two million years ago. As with *Australopithecus*, there are many species of *Homo*, all of them descendants of one ancestral species. When I was a student, we understood the origin of *Homo* as a major event, tied to tool use, more humanlike teeth and diets, and bigger brains than any earlier hominins. We searched for the earliest *Homo*, evaluating every scrap of bone and every tooth to see if it might represent the ancestor of us all. Today, prompted by new evidence from our team and others, scientists are questioning whether the rise of *Homo* really was the revolution that we used to imagine.

Evidence suggests that in its early days, our genus was a bit player. Fossils from other kinds of hominins greatly outnumber those of *Homo* species in Africa. The most successful of these was *Paranthropus*, a lineage with molar and premolar teeth that were giant, as much as twice the size of *Australopithecus* teeth. *Paranthropus boisei* and *Paranthropus robustus* had skulls that sported jaw muscles so enormous that they raised a bony sagittal crest atop their skulls, like gorillas have today. Their massively thick jawbones must have delivered a crushing bite force. Scientists used to assume these species were vegetarians that ate extremely hard foods, but the evidence today shows a different picture. *Boisei*, which lived in East Africa, from Malawi in the south to Ethiopia in the north, seems to have chewed on the edible parts of tough, grassy plants such as papyrus; while *robustus*, found in South Africa, ate a more varied diet including insects, underground tubers and corms, nuts, and possibly meat.

Australopithecus survived for a while, coexisting with early *Homo* species as well. Their diet and habitat shifted as more grassland covered the African continent. The first fossils of one of the latest known australopithecines were discovered by Lee's son, Matthew, at the Malapa dig site, not

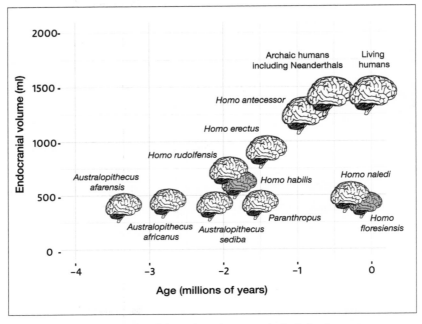

Brain size, calculated from the volume inside skull fossil remains, is a key factor in identifying hominin species.

too far from the Rising Star cave system. This species, *Australopithecus sediba,* lived 1.9 million years ago. It continued the pattern of small body size, small brain size, and bipedal locomotion found in the earlier species—*africanus* and *afarensis*—and it seems to have eaten mostly woodland foods, including the nutritious inner bark of trees, in its increasingly savanna-covered environment. Small jaws and teeth gave it a face more like those found in our own genus, *Homo,* and its pelvis and hands looked closer to human ones as well. In fact, the entire skeleton of *sediba* is closer to the skeleton of *Homo* than that of any other species of *Australopithecus* we know.

The evidence suggests that every one of these hominins could make and use tools. Right now, the earliest stone tools we know about come from Kenya's Lake Turkana region, dated to around 3.3 million years ago. The site where they were found, called Lomekwi, is also known for having fossils of *Kenyanthropus* from the same range of time. Dated to a bit later, 2.9 million years ago, stone tools near Lake Victoria in Kenya were found at a site along with the remains of butchered hippopotamus bone and some of the oldest fossil teeth of *Paranthropus*. More stone tools and animal bones with cut marks dating back to between three million and two million years ago have been found at many sites in Ethiopia, Kenya, and South Africa. South African cave sites have also included tools made of bone, with pointed ends likely used to probe termite mounds or dig up tubers. Paleoanthropologists have studied the hand and wrist bones of *robustus*, *africanus*, and *sediba*, and all of them would have been able to manipulate wood and bone. One can imagine them pounding bones, cracking nuts, and even knapping stone into sharp flakes.

Homo stepped into this scene as another early toolmaker. Scientists have linked a few of the period's fossil fragments from three million to two million years ago to our genus: A part of a jaw from Ledi-Geraru, Ethiopia, represents the earliest *Homo* find, 2.8 million years old. From just before two million years ago, identifiable skulls of *Homo* appear, and from this time onward there were multiple forms of our genus, including the large-brained, large-toothed *Homo rudolfensis*; *Homo erectus*, with its heavy brow and humanlike body shape; and *Homo habilis*, something like a slightly bigger-brained *Australopithecus*.

So far, all the species I've mentioned were discovered as fossils in Africa, but as we move forward in time, we begin to find evidence of *Homo* species elsewhere. The first to disperse from Africa into Eurasia appears to have been *Homo erectus*. The most numerous fossils of its early evolution come from Dmanisi, in the Republic of Georgia, dating back 1.8 million years. A few hundred thousand years later, *erectus* was also in China and Indonesia: Lower sea levels in that age enabled them to walk that far. Nobody is sure why earlier hominins did not travel from Africa into other parts of the world, and we may yet find fossils to prove that they did. But compared with *Australopithecus* and *Paranthropus*, *Homo erectus* had clear advantages over earlier hominins, with longer legs, a human body size and stature, and a cooperative strategy for hunting and gathering plant foods: the fundamental features shared by all our ancient ancestors.

But there was more than one way to succeed in new places. Some species evolved to be very different from *erectus* and expanded the range of hominins even farther. The islands of the Philippines and Flores were never connected by land bridges to the Asian mainland, even when sea levels were at their lowest. Hominins reached Flores before one million years ago, and Luzon, the northern island in the Philippines, by 700,000 years ago. The early evidence is from stone tools. Fossils found in both places date to a period much later—most more recently than 150,000 years ago—and have been designated *Homo floresiensis* and *Homo luzonensis*. Both had very small teeth and bodies much smaller than people have today, especially *floresiensis*, nicknamed "the hobbit." The one skeleton we have representing this species combines rather large feet with a brain much smaller than any found in *Homo erectus*. These species may have evolved from earlier populations of *erectus*, although some scientists suggest *Australopithecus* or *Homo habilis* instead.

Homo erectus survived a very long time, with its latest known fossils, from Indonesia, dating back about 100,000 years. But it did not persist everywhere. In other parts of the world, new species arose. In Africa, early humans with larger brain sizes began to appear just under a million years ago. In Europe, the species *Homo antecessor* appeared by 800,000 years ago. Now DNA evidence can begin to enlighten the story, and it tells us that three ancient human lineages began to separate from one ancestral population around 700,000 years ago. One of those lineages remained in Africa and ultimately evolved into the human species that we recognize today: modern *Homo sapiens*. The other two quickly sprang from a shared pool of founders into Eurasia, one heading west and the other east.

The westerners we call the Neanderthals. Possibly the best known group of the entire early hominin tree, they inhabited the area from Spain to Uzbekistan, Poland to Israel, from more than 450,000 years ago up to around 40,000 years ago.

The easterners are more mysterious. Known as Denisovans, their DNA was first recovered in 2010 from a bone fragment found in Russia's Denisova Cave. This DNA has been identified in only a handful of fossils, none of them skulls or skeletons, so we are still working to understand more about them, including where and when the population lived. A rich record of fossils found in China, dated to between 700,000 and 400,000 years ago, shows a population with brain sizes, brow ridges, and general body shapes similar to those of Neanderthals or archaic African populations, and they may be found to have Denisovan DNA.

These three geographically distinct populations—Neanderthals, Denisovans, and the ancient African humans—evolved genetic differences over the time they existed, but they also exchanged genes with one another. Those genetic exchanges were greatest between Africans and Neanderthals. Denisovans and Neanderthals also mixed, and it appears that both may have received genes from an even more divergent ancient population, maybe *Homo antecessor* or *Homo erectus*. So today, it's no surprise that many people have a small fraction of DNA from these ancient groups. Separation and hybridization were both part of our journey.

That ability to mix suggests that all these ancient people could communicate with one another and could learn other cultures. The archaeological record confirms that they were alike in many of the behaviors that left physical traces—such as tools, animal bones, and campsites. All these kinds of humans used fire, they sustained traditions of finely flaked stone tools, they hunted large mammal prey, and they gathered a wide range of wild plant foods. Sometimes they made and wore ornaments like shell beads, and they used natural pigments such as the red mineral known as ocher. Sure, they looked different, but their minds seem to have been very much alike.

I spent a lot of the first 15 years of my career thinking about how today's people are connected to the ancient branches of African and Eurasian humans. The rise of our own populations—the people that anthropologists call "modern humans"—clearly began in Africa. African people are more genetically diverse than populations anywhere else today, and that diversity first arose some 300,000 years ago. Compared to those in Africa, populations throughout Eurasia and Oceania today are less genetically diverse. These populations all emerged from a founder population that dates back only 100,000 years, and that founding bottleneck is written in their genomes. In the Americas, Polynesia, and some other parts of the world, today's people arose from even later founder populations, as reflected in their genomes.

I believed, as most people do, that our large brain sizes, our unparalleled ability to learn from each other, and our technology all make us the ultimate competitors. How else could we explain our spread into every habitat on the planet? How else could we understand how the diverse tree of our ancient relatives was pruned down to a single branch: our own species? The other forms of humans—Neanderthals, Denisovans, even *erectus*—lasted the longest and, like modern people, had large brains. This was no march of progress; it was a symphony in which dozens of varied melodies emerged, sometimes harmonizing and sometimes discordant, always changing, until at the end a few strains converged into one single note.

This is what I thought before we started working at Rising Star. If somebody had told me that even within their African homeland, our *Homo sapiens* ancestors were not alone, I might have believed it, but I would have assumed it to have involved another kind of human. Maybe it would have had a different-looking skull—the kind of features that get anthropologists excited about new species—but with a brain size, a body size, and a way of life very much like the other humans. I could just imagine that on an island like Flores, a very different species might evolve in isolation without competition from humans. But that should have been impossible in Africa, the very center of our evolution.

What we have found at Rising Star, then, has been a complete surprise: a new species that was not human at all yet was on the stage when our own species, *Homo sapiens,* first arose. *Homo naledi* already existed when the Neanderthals and Denisovans got their start. The *naledi* branch of the hominin family tree may have started up even before the *Homo erectus* branch did. How was all this possible? That's what we had to figure out. The clues would come from our team's journeys into the cave.

I'll let Lee pick up the thread from here.

| 3 |

FINDING *HOMO NALEDI*

Africa is known as the birthplace of humankind for a reason. Our closest living relatives today are African great apes; chimpanzees and bonobos are closest to us in terms of DNA similarities, and the next closest are western and eastern gorillas. And, as John explained in the previous chapter, based on the evidence we have today, the earliest hominins lived and evolved in Africa. Time and again, evidence shows us that critical evolutionary experiments within hominin species arose in Africa before they spread into the rest of the world.

But why Africa?

Africa's sheer size and unique geographic position make it a natural place for hominins to appear and flourish. Africa straddles the Equator, with significant land both north and south of the equatorial line, which provides tremendous biological and geological diversity, from sultry jungles to dry savannas. The continent's position made it immune to Earth's more recent ice ages, which dramatically reduced habitable land and wiped out dozens of species elsewhere. While the rest of the planet recovered from those environmental resets, Africa largely thrived, fueling the evolutionary engine for our distant relatives—and ultimately us, too.

When I began my work in paleoanthropology as a Ph.D. student in the early 1990s, South Africa was reemerging as a hot spot for breakthroughs in evolutionary sciences. In fact, one of my first major finds happened just a few kilometers from Rising Star in Gladysvale Cave, where I discovered a set of fossilized hominin teeth. Gladysvale became the first new site with hominin fossils found in the area in half a century, and even though, over the course of the ensuing decade, few major new discoveries were made here, the region became my passion and my focus.

Seventeen years passed before I became involved with what one might call a major discovery. In 2008 my son, Matthew, then nine years old, was accompanying me on an expedition in the dolomitic region near Gladysvale to explore fossil sites I had recently discovered. He noticed a fossil sticking out of a rock and called my attention to it. I identified it, and a few other bones in the same block of breccia, as hominin. This was that breakthrough I needed. Hominin fossils are incredibly rare, and at the time, such a discovery was considered the find of a lifetime. We named the site Malapa, "my home" in Sesotho, and my team began to explore it.

We eventually revealed two nearly complete skeletons with a combination of characteristics unlike any in other hominins found so far. These hominin skeletons' small skulls and brains resembled those of *Australopithecus*, but their small teeth and jaws seemed more like the ones found in *Homo habilis*, or even *Homo erectus*. The hips of the skeletons seemed quite like ours in some ways, more than the widely flared hips of the Lucy skeleton or of other members of *Australopithecus*. Yet the shoulders and rib cage were built in an apelike way, with long arms for climbing. We were able to determine that the individuals had lived close to two million years ago. Our feeling was that this species might be related to *Homo*, maybe more closely than any other species, but its body seemed made for a lifestyle more like that of *Australopithecus*. In 2010 we announced the

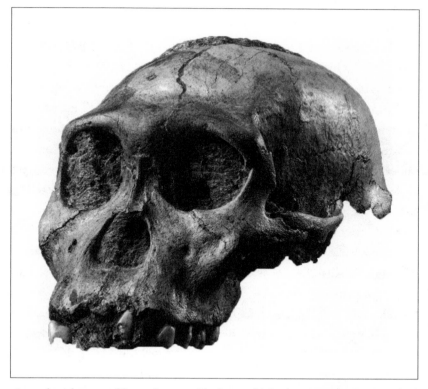

Australopithecus sediba—*discovered by Lee and Matthew Berger in 2008—had features closer to the genus* Homo *than other species of its own genus.*

discovery as a new species, an australopithecine we called "*Homo*-like" and named *Australopithecus sediba*.

The discovery of *sediba* changed my life. It not only allowed me to grow my research program, but it also demonstrated that, contrary to the proclamation of some scientists just a few years earlier that there was nothing new left to discover in South Africa, there were indeed hidden treasures to be found there, like those in Malapa—and often right in front of us.

Within a few years, the *sediba* project grew into one of the largest collaborations ever mounted in paleoanthropology. More than 100 researchers from around the world joined various aspects of the work, concentrating largely on excavations at Malapa and laboratory work analyzing the fossil treasures they revealed. It felt as if we had won the scientific lottery. The discovery of two such complete skeletons was unprecedented in paleoanthropological history.

But then, in 2013, a series of events reignited exploration. While building a protective structure over the Malapa site, effectively ending fieldwork for almost a year, I first enlisted a former student, Pedro Boshoff, and two recreational cavers, Rick Hunter and Steve Tucker, to explore the underground environments of the nearby region. Rick and Steve were perfect for this job: young, wiry in build, and fearless. They took to the task like badgers burrowing through tunnels for a meal. Working with maps of cave entrances I had created during expeditions in the late 1990s, they came back with accounts of little patches of fossil outcrops in deep underground chambers here and there across the region. Nothing particularly spectacular.

Eventually, they began an exploration of Rising Star, a well-mapped cave system near the famous fossil sites of Sterkfontein and Swartkrans. It was a last-ditch effort; after all, Rising Star had been explored hundreds of times by recreational cavers and even by scientists.

In September 2013, Rick and Steve entered Rising Star to test the limits of the map that prior cavers had drawn. They descended 30 meters underground before making their way up a jagged fin of rock and navigating a catwalk of ledges punctuated by cracks that plunged into seeming nothingness. Steve, trying to position himself to take a photograph, wedged himself into one of these cracks, only to find that his feet passed down into a hidden space below. "It goes!" he called out. Rick joined him, and down they went through a near-vertical network of passages through sheer rock. After 12 meters, the route they chose—the one we would eventually call the Chute—spat them out into a fossil-filled chamber, with bones sitting in plain sight on the floor. The two cavers had discovered Dinaledi.

A few nights later, Steve and Pedro arrived at my front gate. On his laptop, Steve showed me images of this chamber, with bones including a hominin jaw with teeth attached and something white and rounded sticking up from the floor. It looked like a hominin skull to me.

It was unlike anything in the history of our science to see ancient hominin remains like this, just lying on the surface, exposed. Steve said that the chamber where they found the bones was extremely hard to get to, but to me it seemed like a treasure trove. Had we really stumbled upon a second miracle discovery?

My team assembled at the Rising Star cave system in early November 2013. The *sediba* fossils had lain close to ground level, but this new setting demanded a new sort of excavation, one unprecedented in African archaeology. Six excavators would descend into the cave complex, supported by a crew of more than 40 others. Team members aboveground managed logistics, tracking people going in and out of the cave and keeping a log of every fossil brought up for study. A paramedic remained on-site in case anyone suffered an injury. Volunteers from the local caving club rigged the cave with power, lighting, and cameras, all designed for remote communication between those of us on the surface and the excavators working in the deep, dangerous spaces. As we retrieved fossils from the site, scientists on our team conserved them and recorded data on the finds. I situated myself in a tent we dubbed the Command Center, where I could watch all that was happening underground on video monitors and discuss the excavation with those team members inside the cave via intercoms and microphones. It was an exciting role, but frustrating at the same time—I wished I could be down in the cave myself, but I didn't even consider it. I knew my body would never fit through those narrow passages, particularly the Chute.

Our first major find was the jawbone that Steve had showed me on his laptop. Once I held it in my hand, I found it was smaller than I had expected from the photos, with rounded, bulbous molars attached to a

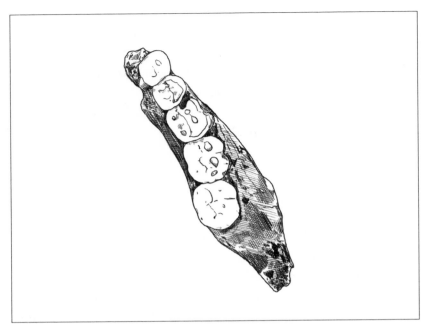

A jawbone with teeth, the first fossil brought out of the Rising Star cave system, was just the beginning of an extraordinary set of discoveries.

thick jaw. These features seemed to match those of *Paranthropus robustus*, an early species found in other South African sites. But the teeth were only slightly larger than our human teeth today, whereas *robustus* teeth were typically twice the size of ours.

We also found pieces of two different right thighbones, which implied that more than one skeleton lay in the chamber. The thighbones had characteristics similar to those of early human relatives like *Australopithecus* and *Paranthropus*, but a cheekbone and a piece of skull we found suggested more recent human relatives, such as *Homo erectus*. The more bones we found, the more the anatomical picture of the findings became twisted and complicated, with features resembling those of many different human relatives from different time periods in the past. And the hand bones we found didn't seem similar to those of any known hominin species, with a long, thin, powerful thumb connected to a narrow, humanlike wrist.

But our biggest find brought the most mysteries of all. As our team worked to excavate a skull fragment from the chamber floor, all the brushing and scraping with paintbrushes and plastic spoons revealed that this skull sat within a complicated array of bones and bone fragments. Working outward from the skull, the team exposed leg bones, arm bones, and bones from hands and feet angled in different directions. We named this tangle of fossils the Puzzle Box. Excavating it felt like a high-stakes version of pick-up sticks, in which each piece had to be carefully extracted without disrupting the others. In total, as the excavation team revealed it, the Puzzle Box grew to an area about the size of a standard suitcase, densely packed with fossil remains.

As mixed up as the Puzzle Box was, some of its contents fell into a sensible arrangement. We found anklebones together with a tibia and fibula of the lower leg, as if they had been connected when the sediment covered them. Two vertebrae and a rib similarly remained articulated—connected as they would have been in a living body—as did parts of a skull and a nearby mandible. A femur jutted almost vertically through most of the deposit, suggesting it was in place before the other bones. This seemed more than a random jumble.

After just three weeks in Dinaledi, we had recovered more than a thousand bones and bone fragments, and aside from a few bones from rodents and an owl we found on the surface, all the fossils appeared to be hominin. It was an unprecedented situation. At most sites in the Cradle of Humankind, hominin bones usually represent a tiny portion of the find compared with the number of bones belonging to other animals.

As our research extended into the next year, up to March 2014, the Puzzle Box pieces started to add up to recognizable anatomical features: an entire head including face, jaw, and skull; a fully articulated hand and wrist, with the fingers curled into the palm; an articulated foot; another partial skull; a young child's jaw with a complete set of teeth; and eggshell-thin fragments from a child's cranium. We had bones of almost every part of the skeleton, from children and adults, as well as every

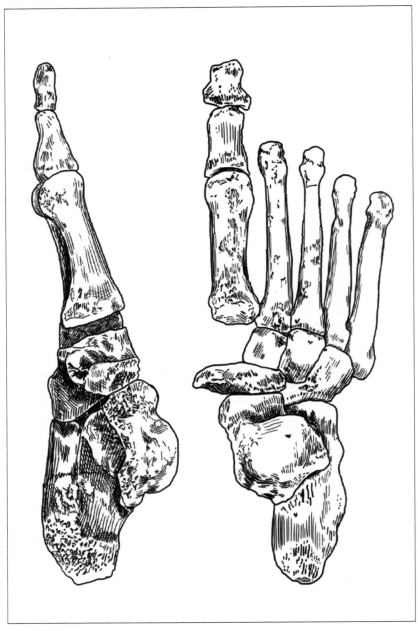

The reconstructed foot of Homo naledi *looks similar to a human foot, differing only in its slightly curved toe bones.*

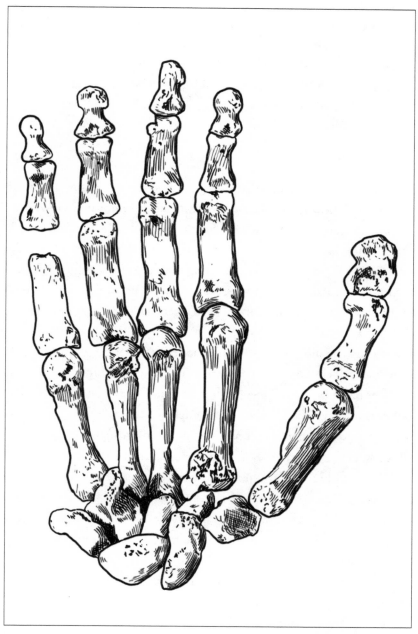

Reconstructed from various fossil finds, the hand of Homo naledi *includes an elongated thumb along with wrist and palm bones similar to those of humans.*

permanent and baby tooth. But we still had little idea what human ancestor we had found. Some parts of the skeleton matched fossils from species at other sites, but all the parts we had found were consistent with one another: All the hands looked alike, all the feet, and all the teeth. Whatever these fossils were, they came from one kind of being.

In May 2014, we assembled a workshop to study the fossils. Scientists from around the world worked to build a detailed picture of this unknown species and compare its fossils with those across the entire hominin record. The skeleton had some humanlike features, but other aspects of its anatomy made it resemble earlier hominin species. If we were going to classify it within the fossil record, we needed to determine whether it belonged in the *Homo* genus, with humans, or in a genus like *Australopithecus* that signified ancient human ancestors.

Based on the more than 100 anklebones and foot bones we had collected, we felt confident that these individuals walked bipedally, like us. The feet had a big toe aligned with the other toes as well as a low arch, as humans have. The leg showed an angle at the knee joint similar to what's found in other bipeds and distinct from primates that walk on all fours, and the overall leg structure suggested the new species could walk and run like modern humans. Based on an assembly of the most complete bones, we envisioned an adult body about the size of some modern small-bodied humans, measuring between 1.3 and 1.6 meters tall (about four feet three inches to five feet two inches) and weighing between 36 and 54 kilograms (80 to 120 pounds).

The Rising Star hominin's hands were mostly humanlike, too. It had a long thumb, wide fingertips, and wrist bones that suggested a strong grip and the ability to manipulate objects. However, its finger bones curved like those of climbing primates, and the first metacarpal bone at the base of its thumb differed significantly from that of both modern humans and

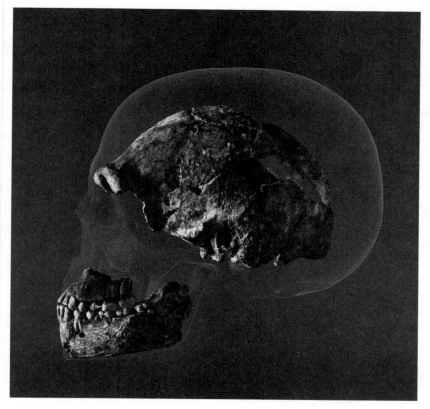

Comparing a composite naledi *skull with a typical human skull shows how small the head and brain of* naledi *were—less than half the size of a modern human's.*

other hominin species we knew. Apparently this species used its hands for climbing *and* activities that required fine motor control, such as tool use.

Skull and tooth fossils tell paleoanthropologists several critical pieces of information about the species they belonged to, namely brain size and diet. The Rising Star hominin's brain and teeth made classification difficult. The skulls we collected ranged from 450 to 550 cubic centimeters in volume, about one-third that of the average human skull, which pointed to a brain smaller than those of *Homo habilis* and *Homo erectus* and most similar to *Australopithecus*. But their teeth—specifically their first molars, canines, and incisors—replicated human teeth in size. That suggested a humanlike diet.

It was a unique combination. The small brain size raised the possibility that these new fossils were some form of *Australopithecus*, like *sediba*, but these new fossils were clearly not *sediba*. While *sediba* was humanlike in the size of its teeth and the form of its hand, this species carried greater similarities to humans in its fine motor structure, feet structure, leg structure, and head shape. Even with its brain size being smaller than normal for the genus *Homo*, the Rising Star hominin had an anatomy suggesting a species that interacted with its environments in a humanlike way.

In the end, we concluded that we had a distinct species in the genus *Homo*, something new to science that walked like us humans, with hands and arms capable of climbing and the fine motor skills characteristic of later toolmaking hominins. It was a strange being, though—relatively tall and skinny compared with other human ancestors, but powerful, judging from its muscle markings and joint sizes. Topping this gangly, long-legged body was a tiny head, with a brain barely larger than a modern chimpanzee's. By assigning this somewhat pin-headed being to the genus *Homo*, we knew we were pushing the limits of what most scientists thought of as the lower brain size allowable in our genus. But the overall body adaptations to the environment pointed in the same direction as those of other *Homo* species, unlike the more apelike adaptations we saw among australopithecines. So the genus *Homo*, we concluded, was where this species belonged.

To distinguish it from other species, the Rising Star hominin needed its own name. We wanted to use a Sesotho word to align the species name with the Rising Star cave system, where it was found, so we decided on *naledi*, meaning "star." And we named the chamber where we had found the fossils Dinaledi, meaning "chamber of many stars."

| 4 |

THE WORLD MEETS *NALEDI*

We first published our findings on *naledi* in the journal *eLife* in September 2015, in an article simply titled "*Homo naledi*, a New Species of the Genus *Homo* From the Dinaledi Chamber, South Africa." The reaction was mixed. Media everywhere reported the story immediately. The scientific community was impressed by the scale of the discovery, and most accepted that the evidence from fossils representing more than 15 individuals was enough to proclaim a new species. But a few scientists spoke out against the designation, arguing that based on the teeth (yet disregarding the rest of the head and skeleton), *naledi* was really just a "primitive" *Homo erectus*. This assertion was driven by the principle that in species diagnosis, only teeth mattered: an outdated mode of thinking that had arisen because of the near absence of other parts of hominin anatomy in the fossil record of Africa. In our field, large numbers of postcranial remains—bones below the skull—were rare, but we had lots of them representing *naledi,* and so we used them.

Others critiqued our interpretation of the assemblage of bones in the Dinaledi Chamber as an example of deliberate body disposal. The great

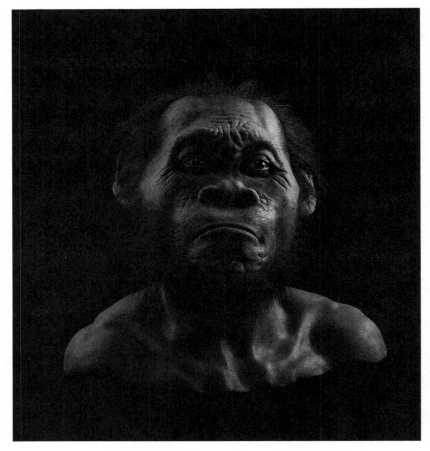

A portrait of Homo naledi, *created by artist John Gurche using information from the fossils found at Rising Star*

majority of scientists held fast to the idea that only large-brained hominins—humans and Neanderthals, in other words—had the mental capacity for ritual practices associated with death. You could almost say it was a foundational principle of our field of study in 2015: Big brains allowed for complex behaviors, like mortuary practices and the recognition of the permanence of death, the assumption went. Since *naledi* had a small brain, it simply could not have had the mental capacity for such behavior.

Some critics even went so far as to suggest we had "ruined" a great discovery of so many fossils by attaching this interpretation of mortuary practices. But our team felt strongly that there was no other conclusion to be reached based on the evidence, and so, as radical as we knew it was to propose that a small-brained hominin had practiced rituals related to death and disposed of the bodies of their kin in this remote chamber, we stuck to our guns. We were, however, extremely careful to use the word "ritual" only to mean a repeated action. We also avoided any language related to burial—no words that might imply that *naledi* had gone into the cave, dug holes, and placed bodies in them. Since we had found no evidence of fire, and since we interpreted the dense assemblage of bones as a layer of remains across the whole subsurface of the chamber, we avoided the controversial term "burial" in every account we gave.

Other critics took issue with our decision to publish the discovery without dating *naledi*. Given the context of the find—on and in the shallow subsurface of the cave floor—there was no way to accurately date these finds at the time. Their anatomy, however, suggested they could be millions of years old. So we decided to publish them without a date, describing this new species based only on its anatomy—a radical decision to many of our colleagues around the world who considered the date of a fossil to be the single most important piece of information about it. Knowing *when* ancient individuals lived is one way that scientists can tie them to ancient environments that may have shaped their evolution, but it's not everything. The anatomy is most important for understanding the place of a species in our evolutionary tree.

Dates are important, though—we all knew that—and so we threw ourselves into the challenge of how to determine when this new species had lived. We began by using the old stalwarts of dating techniques, such as radiocarbon dating—a method we suspected would turn out to be useless, since it can date back only about 50,000 years. The anatomy of *naledi* suggested a species millions of years old, not thousands, but we went ahead and applied radiocarbon dating to some of the fossils, almost out of a sense of duty.

One of the earliest results shocked us. The laboratory carrying out radiocarbon analysis of three *naledi* bone fragments said that two of the fragments were estimated to be less than 35,000 years old. The lab suggested that they were not very confident in this result, because the chemistry of the bone sample looked similar to samples of cremated bone. We didn't know what to make of this result at first. On the one hand, this was getting close to the range of ages in which radiocarbon dating was more likely to reflect contamination from sources of carbon other than the bones themselves. On the other hand, they were results. Could the bones really be younger than anyone expected? We waited to receive the reports based on other dating methods.

In all, six methods were applied to the problem. We wanted to date not only the bones but also the cave context in which they lay. Most of the dating techniques worked only for rock, but at least one—electron spin resonance (ESR)—could be applied directly to hominin teeth. The initial results were somewhat scattered. Results on one *naledi* tooth suggested it was much younger than the others: around 104,000 years old, with a range of error around 29,000 years. This particular tooth was highly worn, though, its enamel so thin that our dating team considered these results less accurate than those from the other two teeth, which had better enamel preservation. The results on those two teeth dated them between 139,000 and 335,000 years old. All in all, we accepted the oldest of these dates—335,000 years—as a maximum age of the fossils.

Another method, uranium-series dating, could date the nearby flowstones in the chamber, potentially providing results beyond the 50,000-year limit of radiocarbon dating. One flowstone a bit higher on the chamber's wall had a bit of sediment adhering to its bottom, including a chunk of bone. Since this flowstone must have formed after the fossil entered the cave, we could use the date of the flowstone to pin down a minimum age for the fossil. Tests dated this flowstone at 243,000 years

THE WORLD MEETS *NALEDI*

Suggestions that Homo naledi *deliberately disposed of their dead inspired this imaginative re-creation of a scene from their lives.*

old, with an error of just under 7,000 years. Hence, the fossil had to be older than 236,000 years. Through all these measurements, we came up with a bracket for the age of *Homo naledi* that we considered as scientifically accurate as the evidence could provide, and we dated the Rising Star fossils back to between 236,000 and 335,000 years before the present day.

That age came as a shock. It not only was surprisingly young, given the anatomy of *naledi,* but it also represented a critical period in the history of Africa and of human evolution in general. That period of time, most paleoanthropologists agree, was exactly when our own species, *Homo sapiens,* first evolved. But most researchers had assumed that humans were alone in Africa at the time they first evolved. Now another species had entered the picture: *Homo naledi.* Was this species capable of some of the advanced behaviors that scientists had attributed only to our own species?

Now we had even more controversial findings to announce. Suddenly, with a date, our research had taken a sharp turn away from anatomical interpretation and into the world of behavior. We needed to know the culture of *naledi,* if it had one. Did it have fire? Was it making and using tools? And if so, what were they?

The more we learned, the more we recognized we had to go back to the cave to try to understand how *naledi* lived.

| 5 |

CHAMBER OF MANY STARS

The ancient rock formations of the Rising Star cave complex date back more than two billion years. The predominant stone in these caves is dolomite banded through with horizontal layers of chert. This combined material is darker, harder, and denser than limestone, which makes it more resistant to the percolation of water that forms cave chambers. As a result, Rising Star has changed slowly. Some spaces have survived for hundreds of thousands of years. While dramatic changes have occurred, caused by a major earthquake or a massive collapse, these events are special occasions—they typically happen only once every tens of thousands of years.

Rising Star's caverns connect through a latticework of crevices and passageways. The best way to envision the system is as a series of abandoned skyscrapers buried in the rock, with tall openings plunging down. The deepest ones might equal the height of the Chrysler Building, in New York City, reaching down some 300 meters. Many are shorter, a half or a third that height. Water falling from one chert-banded floor to the next causes never-ending erosional deconstruction. Everything heads downward with the pull of gravity: rocks, dirt, bones, water.

Horizontal passageways connect these spaces—skywalks, to continue our skyscraper analogy—carved out over millions of years by flowing water, which erodes the limestone more quickly than the chert. Exploration here requires either traversing up and down through carved-out passages or moving laterally through narrow connecting tunnels. It's a complicated labyrinth.

The route down to Dinaledi from the Rising Star cave entrance includes climbs, squeezes, turns, drops, and dangerous leaps over seemingly bottomless pits. After entering the cave system, you first reach an open skylit chamber that we had developed into the Command Center, having moved it from its former location aboveground. Here, we hold briefings for journeys into the rest of the cave system, and explorers kit up by checking batteries, lights, and other equipment before continuing their adventure.

The Skylight Chamber is six or seven meters underground. The sun shines through the branches of a wild olive tree on the surface above, creating dappled light on the stone floor. The outpost here includes plastic tables, chairs, and electronics powered by electricity brought from about a kilometer away. Several chinks in the cave wall have been plugged by rock-filled dead ends. These might have been passageways in the ancient past, but today we can go in only one of two directions: southwest toward Dinaledi or north into a different set of chambers that, as of 2014, we hadn't explored in detail yet.

When you head down the southwest passage toward Dinaledi, you quickly enter darkness. You pass through a series of medium-size chambers, always descending slightly along hard chert floors. Within a few dozen meters, you enter a downward-trending crack that narrows into a three-meter drop. Here, we've placed a permanent ladder to help those carrying gear descend more easily. After the ladder, you scooch sideways until another two-meter drop delivers you into a larger chamber. You're about to face the first serious squeeze of the journey: Superman's Crawl, a body-size hole in the wall barely big enough to fit through. The name Superman's Crawl suggests the position you must assume to traverse it, with one arm in front of you, pulling you forward and pushing your equipment, and the other arm flat against your torso. We've widened the passage slightly to make it more accessible, but it's still a tight fit for most cavers. After seven meters of worming through the squeeze, you emerge in the Dragon's Back Chamber.

You can stand up in the Dragon's Back Chamber, although it won't exactly comfort the claustrophobic: It measures about 25 meters in length and just six or seven in width. A cathedral ceiling towers above, supporting bright stalactites glistening with water droplets. At the far end of the chamber is a huge, jagged fallen block that extends up into the darkness toward the infamous Chute.

This block is the structure for which the Dragon's Back is named: a narrow ridge bordered by precipitous drops on either side. The position of the thin bands of chert running through the dolomite slab hint at an ancient past when the whole structure hung above the chamber floor, a thin wedge suspended like a curtain before it fell as one giant block. Now it rises more than 10 meters from the chamber floor, punctuated by spiny peaks along its ridgeline. For safety's sake, we established a system of bolts, ropes, and harnesses to aid team members up the Dragon's Back. But even with the safety ropes, you don't want to fall. A plunge into the crevices on either side of the ridge would be painful and nearly impossible to escape, if not fatal.

At the top of the Dragon's Back, you have to cross a bridge over a sheer drop to reach the ledge where the Chute entrance waits. This traversal—little more than a long step for most—is far less frightening than the 15-meter downward passage on the far side.

Precautions abound at the top of the Chute. An intercom sends check-ins to the Command Center before anybody attempts the descent, and we perform a secondary gear check to ensure everyone has a full set of working caving gear, including helmet, headlamp, and overalls—requisite in Rising Star. In addition to the gear and comms check, we station a team member at the entrance of the Chute to guide the traveler and report any accidents to Command. We call this person, affectionately, the Chute Troll.

The Chute poses different challenges to different people. The limiting factor for most cavers when it comes to fitting through tight passageways is the expanse of their rib cage: If your ribs can pass through an opening, the rest of you probably can, too, and vice versa. But the Chute houses one of the tightest squeezes most will ever encounter in Rising Star—the cruel opening is just 19 centimeters (about seven and a half inches) across at its widest. Some experienced cavers have mentioned that passing through the Chute is challenging but relatively safe, compared with other maneuvers—being wedged between rocks all the way down makes it hard to slip and fall. But even for those adventurers, the Chute requires painful contortions in response to the unforgiving twists in the rock. The best cavers take almost 10 minutes to travel those 12 meters. Most people take longer, sometimes up to half an hour.

And of course, what goes down must come up: After the reward of spending time inside Dinaledi, everyone has to work their way back up the Chute, an entirely different sort of athletic demand on mind and body. Making the climb depends almost entirely on your upper-body strength, and this time, you have to fight gravity instead of letting it assist you. It's extremely physically demanding, even for the fittest explorer.

Dinaledi's incredible trove of fossils, and the epic effort it takes to reach it, naturally captured a lot of our attention in the wake of the *naledi* discovery. During that 2013 expedition, after hearing me discuss a femur recovered from the Dinaledi Chamber, Steve and Rick had pulled me aside and whispered that they thought there was another such bone in a different side passage, one more than 100 meters away, to the right of the Skylight Chamber. I had told them to keep it quiet until the end of the expedition, when I was certain we had accomplished the task at hand. Now it was time, and so I sent the team into this new and different space to bring back pictures.

The first images Steve and Rick showed me were enticing: I saw parts of leg and arm bones and even fragments of a skull. Like those in Dinaledi, the bones appeared to be simply lying there on the surface, waiting to be excavated. They were clearly hominin. Could they be the same species as the one we had found in the Dinaledi Chamber—the one for which we now had so many bones? That seemed too good to be true.

Then Steve and Rick showed me another image appearing to show chunks of charcoal on the chamber floor. These might not sound as sexy as skull fragments, but for researchers like me, charcoal carries one large, precarious implication: fire. Connecting fire to an ancient species raises a host of gnarled questions that can tangle explorers up in knots, so when someone mused this might be evidence of *naledi* having fire, I dismissed it as insufficient. We needed evidence in context. But the idea did inspire us to name the new chamber Lesedi, after the Sesotho word for "light."

I sat in my usual seat behind the computer monitor at the Command Center as Steve and Rick talked me through their pictures. I listened to them describe the chamber, and how to get there, and as they told the story of their most recent thrilling plunge down into a hidden trove of ancient fossils, I felt a familiar tingle of the call to adventure. I wanted to see the Lesedi Chamber for myself. I was tired of watching the action through a screen.

PART II
SO MANY BONES

| 6 |

INSIDE THE LESEDI CHAMBER

While I had always shaken my head at the possibility of my big body fitting down the Chute, it sounded easier to reach Lesedi. Steve and Rick explained that from the Skylight Chamber, you descend down a narrow slide and then, reaching an opening, you simply scoot along a rock ledge, dodge one drop, and slide a few meters down a tunnel that, fortunately, appears to have been widened by modern-day miners or cavers. It seemed straightforward enough. It sounded like I could do it.

It wasn't until early 2014, though, that I decided to make the journey. We set up a small expedition just for the purpose of going in after the skull fragments. John Hawks worked from the Skylight Chamber, while Steve, Rick, and Alia Gurtov, a Ph.D. student from the University of Wisconsin–Madison, joined me in trying to reach the fossils. From Skylight, we went in the opposite direction from Dinaledi. Over a boulder was a small hole, barely wide enough to slip into. Down we slithered, along a tunnel with a sharp 45-degree turn, loose rocks biting into our legs. Eventually we reached a precarious rock ledge at the end of the passage. After inching along this ledge, I found myself toeing over the edge of a huge hole.

"What's this?" I called ahead, realizing this was the drop they had described.

"That's the Toilet Bowl," Steve replied. He, Rick, and Alia had already stepped around it to continue on the other side. "Don't fall in."

I leaned over to peer into the circular 15-meter plunge. "Why do they call it the Toilet Bowl?" I asked. Something told me it had to do with more than just the shape.

"If you wind up down there, you'll be in deep shit," Steve said.

I took extra time inching my way around the hole. Next, our group navigated a short series of climbs and squeezes until we found ourselves in a small chamber on the cusp of another test—a tight headfirst squeeze. I stared at the small slot in the rock we had to pass through, barely 30 centimeters wide.

"This is the tunnel you mentioned?" I asked Steve and Rick.

"Yes," they said. "Seem all right?"

"Sure," I lied. Honestly, it looked questionable.

Steve and Rick led the way, slipping into the slot headfirst, then jackknifing out of sight like Olympic high-divers. Alia flashed me an encouraging smile, then dove after them. Through the slot, I could hear her grunting and groaning as she squirmed through the rock.

I took a deep breath. The fossils, particularly the skull, lured me onward. I wanted to see the bones in their natural resting place, and I had packed all the tools I would need to lift and document the skull myself.

I slid into the passage headfirst, bending at the waist until my torso was pointed straight down toward the center of the Earth. Instantly, I felt disoriented. I was practically upside down, feeling my way forward—downward—with my hands. Ahead of me, I could see Alia weaving her way through the center of a huge rock split in two. I followed, rolling onto

INSIDE THE LESEDI CHAMBER

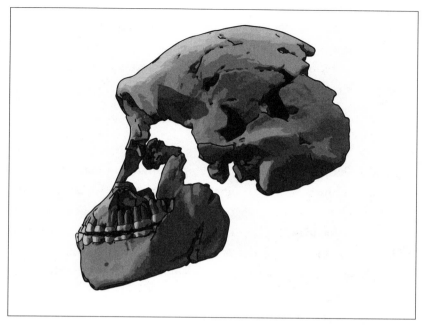

This partial skull, part of a skeleton found tucked away in a stone alcove in the Lesedi Chamber, added a new dimension to the naledi *discoveries.*

my belly, pushing and pulling myself through the same tight space. Finally, after a few meters of rough scrabbling, I was able to stand and join the others in Lesedi.

This chamber had none of the beautiful stalactite formations I had seen in the Dragon's Back or in the images sent up from the Dinaledi Chamber. It was narrow, maybe a meter and a half wide, and shaped like two wedges crossing each other, dead-ending in fissures too small to crawl through. The ceiling wasn't really visible, just a black empty space that continued beyond the reach of my light when I peered upward. In all, it was rather unremarkable.

But then I laid eyes on the bones. I knelt next to the skull, awestruck, tracing its edge with my eyes. The fragment was ovular and the size of a grapefruit. I was mesmerized. For the first time, I was in these ancient beings' space, seeing the bones as they had lain in the dirt for possibly hundreds of thousands of years. My gaze wandered from the skull to the bone fragments scattered beside it. No matter what species this turned out to be, *naledi* or otherwise, Lesedi announced itself immediately as another significant discovery.

Alia and I set up a grid and began taking orientation photos to be used for photogrammetry—the exacting process by which we create virtual 3D reconstructions of every space we study. I could already see that the two-dimensional maps I had seen of this chamber hadn't been able to do it justice: In reality, Lesedi was smaller than my previous impression—seemingly even smaller than Dinaledi—and it was the hub for many passages branching off its main chamber. The Lesedi roof was high, but not as high as ground level.

The bones we found were positioned differently from those in Dinaledi and elsewhere. The skull fragment was in an alcove more than a meter above the floor—a huge contrast to the dense Puzzle Box remains layered in the middle of the Dinaledi Chamber floor. The skull's placement gave the impression that a body had been folded into the wall of the chamber. Excavating the skull took about five hours. We started with the loose pieces on the surface, then extracted the fragments that had collapsed into the cranium. Slowly, a magnificent maxilla—the upper jaw, with teeth intact—began to emerge from the sediment. Ultimately, we exhumed almost a complete skull. I held the short, flat face in my hands and looked at its small, primitive teeth. I had a strong hunch that these bones belonged to *naledi*, as the anatomy looked so familiar.

I packed the fragile fossil remains into padded containers. It was time to make the climb out, which is always harder than the descent. On the way up, your muscles have to strain to push your body upward, and straining muscles are bigger than slack ones. That was a particular problem for my body type. I let Steve, Rick, and Alia go up first. Alia took the skull.

INSIDE THE LESEDI CHAMBER

After Alia reached the entrance again, I heard her shout, "All clear!" I shone my headlamp into the fissure before me and stretched both arms over my head. Then, I pushed off with my legs and squeezed my body into the passage. I felt the space press in on me. I needed to contort into a 90-degree angle for the next turn, following the curves of the rocky wormhole. My arms were almost useless. This was a lot harder than sliding down headfirst.

I wedged my torso through the 90-degree turn, then instinctively tried to bring my legs up to push off with my feet. But there was no room. I couldn't bend my legs, because my thighbone was too long for me to position my legs for leverage. Meanwhile, my arms had no way to pull my lower body up as dead weight. A stab of horror pierced my brain: Maybe my body just couldn't navigate the way out of this hole.

I wriggled and twisted. I wrenched and turned. I squeezed and shimmied. Nothing. I struggled like that for 40 full minutes, exhausted, and still stuck. No matter what I did, I could not contort my torso or position my legs in a way to wedge myself up. I was running out of strength.

"Try rolling onto your back," the others called down to me. No use.

"Do you think you could reverse yourself and come up feet first?"

That acrobatic move proved impossible, too. I had no leverage. My thighbone was too long to make the 90-degree bend. I was getting tired.

My teammates, up above me, outside the chamber, seemed to be tiring, too. Their voices grew softer. They seemed to be talking among themselves.

"What's going on?" I yelled. I tried to sound calm, but I was beginning to feel deeply worried about getting out.

"We have another idea," Rick called.

"What is it?"

I heard him scrambling at the entrance. "We're going to pull you up." A moment later, Rick had climbed back down into the passage and wedged himself into a spot in the passage above me. I saw what he intended and squirmed out of my jumpsuit, leaving me in nothing but a T-shirt and cycling leggings: anything to reduce drag. Rick dangled a nylon climbing rope down toward me, and I tied a double-column knot I'd learned in the Boy Scouts around both wrists, cinching it as tight as I could. The rest of the crew held tight to the other end of the rope. I called out a countdown: "In three ... two ... one!"

The team yanked. Pain shot through my shoulders as my arms wrenched over my head, my body in a diving position. The rocky sides of the wall ripped my T-shirt, but slowly, I began to inch upward. I wiggled, to help in whatever way I could, and I finally felt my lower body angle into the slot. My thighs had made it through! Now I could push with my legs. Rick untied my wrists and climbed back out, and soon I was grabbing that rope with both hands and clambering my feet on the walls of the tunnel as my team members pulled to assist me.

And then I was out.

I stood there, covered in mud but euphoric. It was great to be free, but even better, we now had a spectacular skull to study. Relief flooded my body. We had accomplished what we had gone into the darkness to do, but I decided then and there that I was never going back into that space again. My place was aboveground.

But unfortunately, my team made sure that part of my legacy stayed down in Lesedi. They began calling the 90-degree turn that had trapped me the Berger Box—it even appeared on our 2017 published map of the Lesedi Chamber in the journal *eLife*. It made me laugh. I've always wanted something named after me.

| 7 |

CAVE DWELLERS

The Lesedi fossils produced much of a skeleton, one we could compare to the Dinaledi fossils point by point. In almost every detail, the fossils matched those of *Homo naledi,* and with the findings all together, we could calculate the dimensions of a single *naledi* individual for the first time. We named the skeleton Neo—"gift" in Sesotho—for its essential contribution to our understanding of this new hominin species.

The Lesedi discovery shook up the way we were thinking about the entire Rising Star cave system. Other cave systems in the Cradle of Humankind region contained multiple deposits of fossils, but these invariably represented different time frames and different species. The best known example is Sterkfontein, just two kilometers from Rising Star. There, the largest breccia deposit, now exposed on the surface, contains bones of *Australopithecus africanus* dating back more than two million years. In the same cave system, one deposit contains tools and a few fossils of *Homo* less than a million years old. Finding fossil bones in a deep cave setting was not so strange—although the forbidding entry into Dinaledi raised questions about how they got there. But finding similar bones in a chamber more than 100 meters away: It was rare, and it just did not seem like mere coincidence. It made us wonder: Could *naledi* have been moving through the larger system, cave to cave, and actually using these remote spaces? It seemed hard to imagine. We had thought that hominins shied away from the deepest subterranean spaces, and we had always assumed that the bones we found in such places represented either unlucky individuals who had fallen in or bones that had been washed in, swept by natural causes.

The Lesedi discovery raised so many new questions: If these fossil deposits had been made by the same species, why were we finding the bones in different orientations and circumstances? Why were some of the bones in Lesedi found in alcoves above the floor of the cave, while the Dinaledi bones were either on or within the floor itself?

We came up with two possible explanations: First, we theorized that the floor of the Lesedi chamber had been higher in the time of *naledi* than it is today. Perhaps the Neo skeleton and the other bones were part of a larger deposit, and the ancient floor had eroded away, washing out other bones and leaving these behind. Our excavation wasn't extensive enough yet to disprove this scenario.

The second explanation was more provocative. Perhaps the Neo skeleton had been preserved in the raised alcove because it had been deliberately placed there by *naledi*. Perhaps Lesedi and Dinaledi represented two variations of what might be described as an intentional *naledi* body disposal.

In our first scientific paper on *Homo naledi*, we had carefully suggested that *naledi* might have deliberately placed the bones in Dinaledi. Deliberate disposal into underground chambers is one of many mortuary practices—a term used to describe consistent and purposeful ways of dealing with the dead. The very idea had evoked skepticism among many archaeologists, though, who saw mortuary practices as uniquely human behaviors.

The oldest certain instances of burial date back about 100,000 years, and the archaeological record of humans does not reveal any commonplace mortuary practices until 35,000 years ago. In paleoanthropological

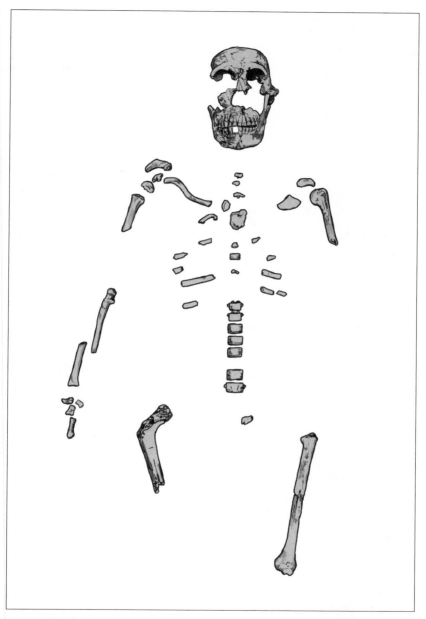

Over the long run, fossils representing much of a skeleton were found in the Lesedi Chamber. Team members named the figure Neo—"gift" in the Sesotho language.

terms, that's yesterday. Here, we were dealing with a hominin species that likely lived at the dawn of humanity, more than 230,000 years ago, with a brain just larger than a chimpanzee's—and we were considering the possibility that it conducted deliberate body disposal.

Now, with what we were finding in Lesedi, we faced even more complexity. Lesedi seemed so similar to Dinaledi, yet in Lesedi, the Neo skeleton looked as if it might have been placed in that alcove in the wall. Was it possible that this ancient species was really going deep into this cave system with the bodies of friends and loved ones and making some kind of distinction between different parts of the cave? If so, it meant that this kind of ritualistic behavior was not just the product of a large human-size brain. Phenomena that had long been deemed uniquely human might be found among *naledi*, too.

| 8 |

ANOTHER BODY

Our 2017 expedition began with one urgent question in mind: How did the bones end up in the Dinaledi Chamber? Most scientists outside our team assumed the least controversial explanation: The bones had tumbled down through the Chute or a similar opening from the Dragon's Back Chamber, one easier to reach, and into Dinaledi. But many on our team, including myself, didn't think that was likely, given the position of the Dinaledi remains relative to the Chute entrance. Our maps supported this theory: They showed the floor of Dinaledi sloping down from the Chute exit toward the deepest parts of the chamber. We called the area near the Chute the Landing Zone. From where the Chute spat you out, the map showed, you walked along a short, narrow passage to the rest of the Dinaledi Chamber.

This first map of the Rising Star cave system was an amalgamation of data from decades of exploration, but it originally was missing a depiction of the Dinaledi Chamber. Before Steve and Rick entered Dinaledi in 2013, there was no known record of the chamber anywhere. We added it to the map based solely on their drawings, tape measurements, and memories.

Our expedition had access to powerful, but limited, mapping technology. Laser scans allowed us to produce a high-definition 3D virtual model of the route to Dinaledi, but the simulation ended at the Chute. The passage was too narrow for the mapping instruments to go any farther into the system. That left mapping Dinaledi up to our explorers' ability to draw and describe their experiences. But the more time they spent in Dinaledi, the more we had to change the map.

Near the start of our 2017 expedition, Steve showed me the most updated map of Dinaledi on his laptop. The latest iteration contained the different levels of the chamber and, for the first time, displayed them in color. But there was a problem. Instead of depicting one short, narrow passage connecting the Landing Zone to the rest of Dinaledi, the new map showed two long passages, each a half meter wide at most, connecting the Landing Zone and the main chamber in parallel. "We've been thinking about this all wrong," I said. "The Dinaledi Chamber isn't one chamber, it's two."

Steve squinted at the map. "I see what you're saying," he said, "but it's still part of the same space. The floor is the same, and there were bones here"—he pointed to the dig area farthest from the Chute—"and here." He pointed to another site near the Landing Zone.

I scratched my head. Steve had been in Dinaledi; I hadn't. But this seemed like an important distinction. The division between the passages created a natural choke point that would impede the tumble of bones down the Chute into the deepest depths of Dinaledi. So how had the bones reached the far part of the chamber? "If bones were sliding down from the Chute, they would pile up at the entrance to these passages, and they would be visible to us unless they've been covered with dirt. They would be scattered on the passage floors, or just under the surface, right?" I asked Steve, pointing on his map.

"Right," he said. "We can test for that."

"Can you dig at the beginning of the two passages?" I asked. "If there's a pile of bones blocked up there, it's a choke point, but if not ..." I paused. "Then the bones came in through some other way."

"We'll try it," Steve said.

I stared at the screen in front of me, once again frustrated at not being able to enter Dinaledi myself and having to rely on pixelated broadcast feeds, maps, and smartphone images. What else were we missing? What else were our pictures getting wrong?

I tried to visualize the spaces in my mind. The long, straight journey down the Chute, the drop into the first open cavern—the Landing Zone—then the twin passages leading to another space entirely. Those passages were five or six meters in length, maybe a bit more—quite a distance. If the bones hadn't tumbled through the passages, had *naledi* crawled through and died in the Dinaledi Chamber? That didn't satisfy me. It would have been possible for other animals to do the same, and there was zero evidence of that. It really seemed like *naledi* habitually entered these parts of the cave. We had to treat it as if they did: as a separate space unique from what we were calling the Landing Zone.

And so the Landing Zone became the Hill Antechamber, a large sloping area at the base of the Chute, separated by two passages from the Dinaledi Chamber and named after Lyda Hill, one of the patrons of our work. We began excavating the space right away.

Our digging in the Hill Antechamber started in two places: one at the top of the slope, against the wall, and the other near the passageway to the Dinaledi Chamber. Soon enough, fossils began to emerge near the tunnels: a few fragments that might be from a hominin tibia, then a few teeth. That handful of teeth begot more teeth, worn adult teeth with small facets that showed they fit perfectly together. We hypothesized that we were finding the front of a *naledi* skull, mostly degraded. There were many long bone fragments as well, but we needed to compare them with the findings at the other site in Hill to figure out whether they began here or slid down from higher up on the slope.

Fragments appeared at the higher site. Within a few centimeters, the team encountered a mass of white powdery material—bone, flowstone, or some combination of both. Using paintbrushes, the team continued brushing away sediment all the way to the edges of the excavation square, exposing a growing patch of chalky white material interspersed with clay. When teeth appeared, arranged as if they were still part of a jaw, another theory materialized: This, too, was a skull—the white powder was the crushed remains of the bone.

The caving team stretched a ladder horizontally across the excavation site so they could rest suspended over the dig area and work without touching anything. They began to dig deeper around the edges of the square. Hand and wrist bones came into view, apparently articulated. Then three or four ribs emerged from the sediment. We had found a skeleton.

This wasn't what we had expected. When we imagined *naledi* to be depositing bodies or bones down the Chute, we came to anticipate bones in a random jumble, like parts of the Puzzle Box. But this was not a hodge-podge pile of bodies or bones. It looked like a single intact skeleton. Some stray fragments of bone appeared to be scattered nearby, but nothing like a dense mound of them. To make matters more intriguing, the sides of our excavation square showed layers of sediment sloping parallel to the floor surface, built up slowly over time. But the angle of the skeletal remains didn't perfectly match this buildup. If a body had been left to decompose on this steep slope, how had it managed to stay where we found it?

We couldn't excavate this skeleton the way we had those in the Puzzle Box. These bones were too fragile; they'd disintegrate if we tried to move them, and they were most valuable in the exact orientation and context in which we had found them. That big picture might tell us about how they ended up on this slope. But the only way to conserve all that information was to remove the entire skeleton from the cave system at once, without disturbing it, and bring it to a laboratory for conservation and analysis. We had handled fragile bones before. But this was a challenge on a different scale entirely. This was going to take some planning.

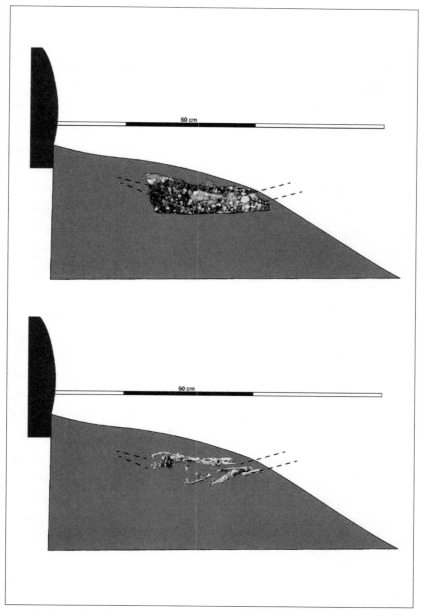

The fossils found in the Hill Antechamber lay at an angle different from the sediment on top, adding evidence for the deliberate burial of three individuals here.

As we neared the end of September 2017, bits of the Hill skeleton still lay exposed. We had decided, in general, that we would excavate around the skeleton so that it was suspended in a pillar of sediment and then cast this pillar in plaster to protect it. From there, we'd simply lift the cast through the Chute and haul it out of Rising Star. An easy job—on paper.

Casting the entire skeleton in a protective plaster jacket presented a problem because of the skeleton's size. Anything that passed through the Chute had to fit through that 19-centimeter squeeze, and although the skeleton was shorter in length than the average human, it was still going to be tight. We spent months taking measurements, looking at projections, and continuing to excavate the skeleton. By March 2018, we had our plan. It was risky.

After we isolated the skeleton, we found that it ultimately measured almost a meter in height and about 30 centimeters in width—too big to push up the Chute. But we found an area running through the middle of the pillar that showed little evidence of bone, so working slowly and carefully, we sifted sediment out of this section until we had separated the pillar into two pieces. It's hard to describe the emotions you feel watching your team members perform surgery on an ancient skeleton via computer monitor, but if I could venture a comparison, I would say it's like watching your team members perform surgery on one of your children via computer monitor. It's a tense moment.

But once the team had created a deep groove between the two pieces, they wrapped the smaller piece in plastic, then jacketed it with plaster. After the plaster hardened, the team severed the piece from the floor and plastered that surface to create a fully enclosed block. On the floor beneath the block, I noticed a small circle of orange clay in the sediment. It struck me as unusual at the time, but there was a skeleton nearby on life support, so I quickly forgot about it. The next day, our team took the severed block out of the cave.

The team cut the remaining block into two more pieces—be still, my beating heart—and brought the smaller of the pair out of the cave. For the last piece, the biggest, it took four team members to coat it in plastic, then plaster, then wrap all that in a waterproof duffel bag attached to the rope we used to haul things up from the chamber. Everyone in Dinaledi that day, scientists and explorers together, positioned themselves above and below the bag to pull, guide, and coax it through the Chute's twists and turns. It was by far the largest and heaviest object we had ever taken through the passage. After more than an hour of repositioning, re-angling, and reorienting the block to negotiate the route, our precious cargo reached the Dragon's Back Chamber intact.

Then came the fun part: studying it.

| 9 |

HINTS OF BURIAL

In November 2018, more than six months after we had removed the skeleton from the Hill Antechamber, I was sitting in the Skylight Chamber's Command Center, supervising two new excavations near the Dinaledi Puzzle Box, when something on the monitor caught my eye. I nearly gasped. "I think I should have them stop the excavation," I said.

Keneiloe Molopyane moved next to me and narrowed her eyes at the computer screen. Kene was one of the new generation of explorers working in the Rising Star cave system. We called her Bones for her expertise in interpreting subtle evidence in skeletons. Her combined training as an archaeologist and a forensic scientist gave her a unique perspective for solving problems. On the screen, we could see the back of the excavators' helmets and a small area of the chamber floor. The light from the excavators' headlamps darted around the cave. "Why stop?" she asked.

Our second 2018 expedition was meant to test whether different parts of the Dinaledi Chamber had the sort of continuous, dense layer of bones that would indicate a natural flow of fossils into it. We had dug two new excavation squares to search for these bones: one south of the Puzzle Box in the direction of the Hill Antechamber, where we had removed the plaster cast that spring; and one north of the Puzzle Box, close to the cave wall.

The square to the south eliminated the possibility of a chamber-wide bone bed almost as soon as we began digging. We found some bone fragments, but as we dug deeper down and farther away from the Puzzle Box, we unearthed fewer and fewer fossils. It was as if bones from the Puzzle Box had spilled out in this direction.

The northern square, however, revealed a heavy concentration of fragments—these bits seemed associated, as if they had come from one individual. It seemed that we had just exposed the edge of a larger trove, so we opened another excavation square adjacent to this one. That space contained half a jawbone containing five intact teeth. Then a large femur shaft emerged, positioned as if it were spiking the earth. As we removed sediment one plastic spoonful at a time, we uncovered a concentration of bones stretching 60 centimeters by 30 centimeters—about the size of an airplane carry-on bag. We dug eagerly around the cluster, but oddly, the surrounding sediment contained only a few bone fragments—and in some places, no bones at all. Between this concentration of bones and the Puzzle Box, we found very few to no fossils. It didn't make sense. If the bones had flowed into the chamber, why had the fossils clustered in such concentrated groups? Why was there an empty space between them? It was as if the excavators had come upon a wall, a distinct vertical division in the sediment.

"It looks like there's a hole in the floor of the cave," I told Kene. "The sediment we're digging through looks like it's been disturbed. I don't think it's a natural depression." The sediments in these caves formed through a slow accumulation of dust and debris coming off the walls of the cave and blanketing the floor in even, consistent layers. But the sediment we were scooping out with our plastic spoons didn't have that same level of consistency. "It looks a lot like a burial feature to me," I concluded.

Bones's eyes widened. "It does," she agreed.

"I'm going to stop them before they disturb something." I clicked on the intercom mic and called to the excavators. I saw them straighten up from their work and look toward the camera. "What's up?" they asked.

"I think I'm seeing something," I said. "I'd like to show it to you." I described what I saw, explaining that from the camera angle, it looked as if the new concentration of bones matched an oval space where the sediment appeared disturbed. It looked like the edge of a hole that had been dug a long, long time ago.

The excavators glanced at each other. These were the two most experienced excavators on the team, and they had been digging out this square for days. They knew what they were doing, and they were there in person, while I was interpreting the geology on a computer screen.

"We don't agree," they said, almost in unison.

Bones shrugged at me. I released the intercom button so that she and I could have an aside. "I am certain we are looking at a hole that has been dug into the floor of the cave," I said. "I think that's why these bones are all together, and I think we need to pause our work until we figure this out."

Bones looked at the image on the screen again. "I think you're making the right decision," she said. "We should stop."

I pressed the intercom button. "I'm sorry," I told the excavators, "but I've made the decision. We are stopping the excavation until we can clarify what we're dealing with here."

I could see the disappointment in their body language. They gathered up their tools and set up to take the day's final photographs.

For five years, we had worked in Rising Star knowing that *naledi* had occupied these spaces. Within days of our first explorations, we had reason to suspect that *naledi* had used the Dinaledi Chamber as a repository for their remains, and nothing we had found since then had disproved that hypothesis. In fact, the accumulated evidence ruled out many alternative hypotheses. But deliberate body disposal—the language we had all carefully used to this point—is very different from burial. Body disposal means bodies are just placed or tossed into a space—literally disposed of—but burial means a body is deliberately interred and then covered. That means there was not only an understanding that death is permanent, a concept typically attributed only to the human mind, but also a ritual, taught and learned, as to how a burial takes place. It's sophisticated.

Archaeologists have found surprisingly little evidence of burial among the earliest members of our species, *Homo sapiens*. The oldest clear cases were found in Israel; they're believed to be between 120,000 and 90,000 years old. In Africa, the oldest known human burial is an 80,000-year-old skeleton, a child, found in the Panga ya Saidi cave in coastal Kenya. Neanderthals also sometimes buried the dead, although the best evidence of buried Neanderthal individuals comes from fairly late in their existence—far less than 100,000 years ago.

Burial takes planning, a shared intention among a social group. Most other kinds of animals treat a dead member of their own species with indifference. A dead individual is simply one that ceases to move along when the group does. Occasionally, some animals have been observed trying to revive a dead member of their family or carrying dead infants about for long periods of time, as if they might come back to life. This has led some archaeologists to consider the recognition that death is permanent, and that we will all eventually die, as central to the very definition of being human. So most scientists have long considered burial as an exclusively human thing, one of the by-products of a big brain, along with art and symbols and language. For them, burials and awareness of self-mortality separate us from all other animals, and indeed separate us from all other human ancestors.

So it was a radical idea to propose that the Rising Star cave system might contain a burial site. We would need extraordinary evidence before we could claim we had found a deliberate *naledi* burial.

As I sat next to Kene and watched the excavators in Dinaledi pack up, I felt as if our work was entering a different conceptual realm. As much as I trusted our scientific methods, there might be dimensions those methods couldn't touch. If this turned out to be a burial, it meant this chamber had some sort of abstract meaning to *naledi*. But science can't observe, measure, or understand meaning. The best we could do was to record and document every bit of physical evidence, even the most minute, to help us re-create the experience of the ancient past. But we had plenty of bones already—more bones than most in our field could dream of finding—so our research goal now went beyond skeletons. We were reaching for something harder to grasp.

The possibility that we were uncovering *naledi* burials was growing stronger. By 2019, we had *naledi* fossils from many different areas in the Rising Star cave system, including the Puzzle Box; the Dinaledi Chamber; the Lesedi Chamber, more than 100 meters away from Dinaledi; and the Hill Antechamber. It was a wealth of geological data, but the differing contexts for the data complicated any cohesive picture we attempted to draw. Our struggle to date the fossils meant we couldn't tell whether the different parts of the cave system were used all at once or at different times, so we could not assume that the *naledi* individuals we were finding represented related events. Even within the Dinaledi Chamber itself, we couldn't prove that the various *naledi* remains we found were within 100,000 years of each other, in the time range between about 230,000 and 330,000 years ago. Neither the situation nor methods available at the time allowed greater precision. Our dating methods simply weren't accurate enough on the whole assemblage, so all we could assert was a bracket of ages.

But we hoped our discovered skeletons might bring clarity to our muddled understanding of *naledi*. We suspected the unexcavated feature in Dinaledi—the possible grave—was a buried skeleton, and of course we had the Hill specimen we had lifted out through the Chute. In the context of Dinaledi, we were considering the Lesedi Chamber as a possible burial

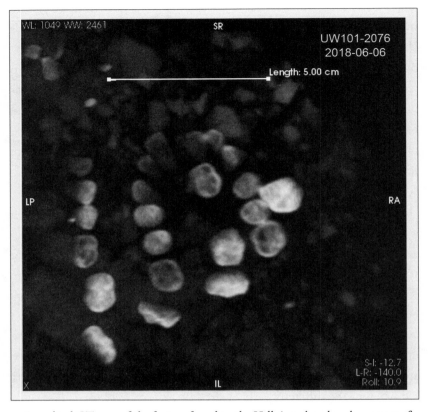

A medical CT scan of the feature found in the Hill Antechamber shows a set of child's teeth in the position they would sit within a skull, though the surrounding bone has dissolved away.

site as well, due to the Neo skeleton's location in the alcove above the chamber floor.

The Hill skeleton represented the most promising evidence of a *naledi* burial. The plaster jacket containing the body had been sitting in my laboratory at the University of the Witwatersrand in Johannesburg since 2018, but it was finally time for it to take center stage.

First, we scanned the two smallest pieces of the block using a microCT scanner, a device that uses x-rays to "slice" the block into images less than one-tenth of a millimeter thick. After the scan, a program combines the slices into a high-resolution 3D picture. The smaller block contained only a few bone fragments, but the larger one contained part of a femur and tibia, a pelvic bone, and either finger or toe bones. It was clearly part of a leg.

The microCT scanner could not accommodate the biggest of the three blocks, so I pulled some strings with the help of my wife, Jackie, a radiologist at the Charlotte Maxeke Johannesburg Academic Hospital. She arranged for us to use the medical CT scanner at the hospital, and though the larger machine scans "only" in slices five-tenths of a millimeter thick, the imaging still revealed plenty.

First, a set of teeth showed up, bright white, all in order, and arched as if they still were inside an upper jaw. Several lower teeth were bunched nearby. Moving down the scan, we saw a jumble of hundreds of bone fragments: ribs, arms, and what might have been hand bones. Moving further, we found five articulated toes. It was a promising initial search, but in truth, CT scanning has its limits for collecting fossil data; it's designed for medical technologies, so paleoanthropologists have to spend a lot of time segmenting the scans manually to delineate every scrap of fossil that the broad strokes of the process might miss. I passed the responsibility over to John.

In January 2020 I was in South Africa, preparing to give a presentation at the European Synchrotron Radiation Facility (ESRF) in Grenoble, France—the world leader in microscale 3D scanning of fossil remains. They use one of the world's most powerful x-ray beams to create incredibly high-resolution views of the inside of fossils, bones, and teeth. It was perfect for studying the Hill skeleton, and I thought the ESRF was our ticket to figuring out what was inside the block without cracking it open. But then John showed up at my office, freshly arrived from the United States. "Come take a look at this," he said, starting up his laptop. I looked over his shoulder as an image of bones appeared on the screen.

John explained that over the past several months, he had segmented the slices of the Hill Antechamber block—more than 2,000 in total—and marked every bone and tooth he could distinguish. What I was about to see was the best visual representation of everything contained inside the block.

With a few keystrokes, John called up the data, and a nearly complete skeleton came into view. I recognized the teeth and the foot from the CT scan, but John's work had brought forth new details—stunning new details. There were body bones here, and some bones had been rendered clearly enough to show epiphyses, the unfused rounded ends of long bones that usually give way to fully developed bone as the body matures. But if they were still here on this specimen, it meant ...

John caught my eye. "It's a nearly articulated child skeleton," he said.

I had never seen anything like those images in my life. It was a child's body, almost certainly of *naledi*, curled up in a space smaller than a laundry basket.

"My first priority was to see if there is some kind of logical arrangement to the bones," John said. "You see that foot at the bottom?"

It was the same foot that we had seen in the slices. Several splintery fragments lay above it, with another bone attached above those. "Are those the leg bones?" I asked.

"The tibia and fibula, both shattered," John confirmed. He pointed to the tallest bone. "I think this is part of the femur."

I visualized the skeleton's full anatomy as he spoke. It was as if the child was curled up in something like the fetal position. "I think these clumps of sediment under the foot are the base of a hole," John continued. "The clumps form a concave layer. Remember when they sliced through the bottom of this block in the cave to jacket it, and we found an orange circle of clay on the bottom? That was the outline of a hole that was dug in Dinaledi."

I looked at the image closely. Just as John had described it, I saw a distinctive dark line on the CT scan, a clear sedimentary change. This

ABOVE: *The Cradle of Humankind near Malapa, the cave where Lee Berger's team discovered fossils of* Australopithecus sediba

LEFT: *Excavation team member Marina Elliott in the Skylight Chamber, near the main entrance of the Rising Star cave system*

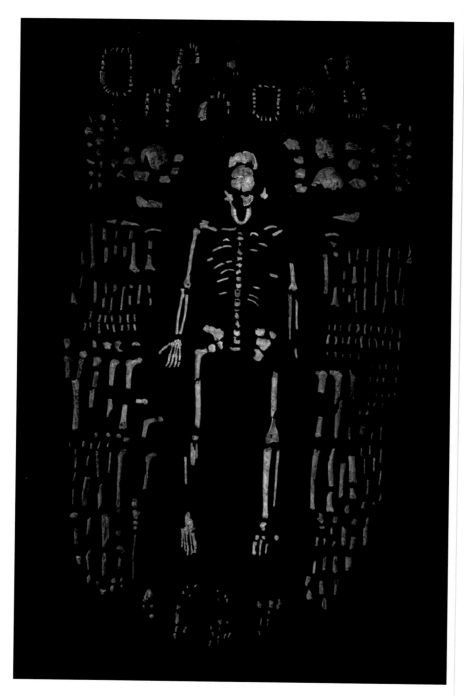

These fossil bones from the Dinaledi Chamber, sourced from numerous individuals, represent nearly every part of the body of Homo naledi.

Artist John Gurche's interpretation of how a naledi *individual may have looked. The patterns of hair and skin color are speculative, while the shape of the face is informed by the anatomy of humans, great apes, and* naledi *fossil bones.*

ABOVE: *Agustín Fuentes (left) and John Hawks (right) supervise the 2022 Dragon's Back Chamber expedition from the Command Center.* BELOW: *Lee Berger (left), Dirk van Rooyen (center), and Mathabela Tsikoane (right) review the expedition plan.*

ABOVE: *Team members Ginika Ramsawak (left) and Erica Noble (right) work on excavation squares in the Dragon's Back Chamber.* BELOW: *Marina Elliott, Ashley Kruger, and Dirk van Rooyen (top to bottom) survey progress in the Hill Antechamber.*

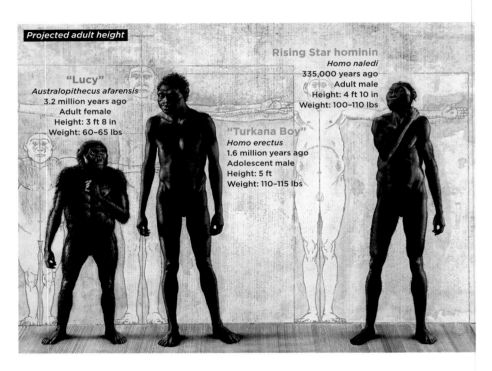

ABOVE: *Compared with* Australopithecus afarensis *(left) and* Homo erectus *(center), Homo naledi had a humanlike stance and body, but its brain and shoulders were more like those of Australopithecus.* BELOW: *Features of the* naledi *foot suggest a humanlike bipedal gait, and a* naledi *hand reveals a mixture of characteristics: broad fingertips, a powerful thumb, a wrist suggesting tool use, and curved fingers suggesting climbing.* OPPOSITE: *Details of the* Homo naledi *skeleton*

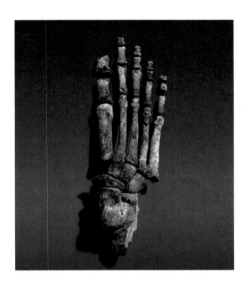

The Sum of Its Parts

A composite skeleton reveals *H. naledi*'s overall body plan. Its shoulders, hips, and torso hark back to earlier ancestors, while its lower body shows more humanlike adaptations. The skull and teeth show a mix of traits.

HOMO FEATURES

Humanesque skull
The general shape of *H. naledi*'s skull is advanced, though the braincase is less than half that of a modern human's.

Versatile hands
H. naledi's palms, wrists, and thumbs are humanlike, suggesting tool use.

Long legs
The leg bones are long and slender and have the strong muscle attachments characteristic of a modern bipedal gait.

Humanlike feet
Except for the slightly curved toes, *H. naledi*'s feet are nearly indistinguishable from ours, with arches that suggest an efficient long-distance stride.

AUSTRALOPITHECINE FEATURES

Primitive shoulders
H. naledi's shoulders are positioned in a way that would have helped with climbing and hanging.

Flared pelvis
The hip bones of *H. naledi* flare outward—a primitive trait—and are shorter front to back than those of modern humans.

Curved fingers
Long, curved fingers, useful for climbing in trees, could be a trait retained from a more apelike ancestor.

ABOVE: *The first public announcement of* Homo naledi *in South Africa. Cyril Ramaphosa, then deputy president, holds a reproduction of a* naledi *skull.* BELOW: *The Rising Star expedition team celebrates after bringing the first fossil skull out of the Dinaledi Chamber, November 2013.*

ABOVE: *Lee Berger struggles out of the Chute into the Dragon's Back Chamber, summer 2022.*
BELOW: *Excavator Keneiloe Molopyane—nicknamed Bones—inspects the ongoing excavation in the Dragon's Back Chamber.*

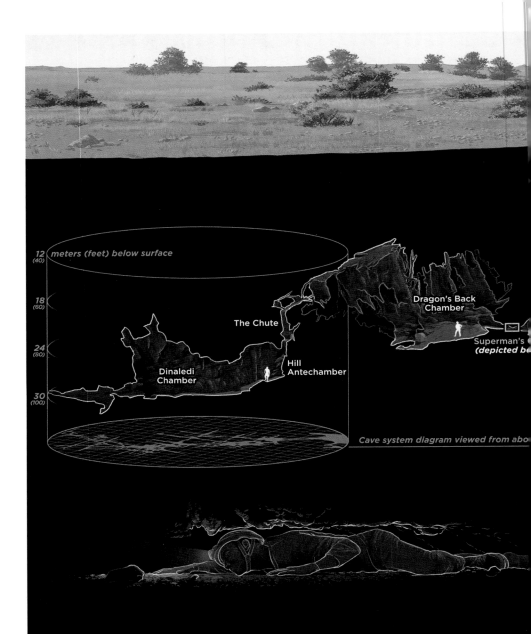

The path taken by Lee Berger and his fellow explorers through the Rising Star cave system to the Dinaledi Chamber is long and arduous. It includes both open chambers and narrow passageways, such as Superman's Crawl and the Chute. These spaces would have posed a different challenge for Homo naledi than for us today, given their anatomy, shown opposite.

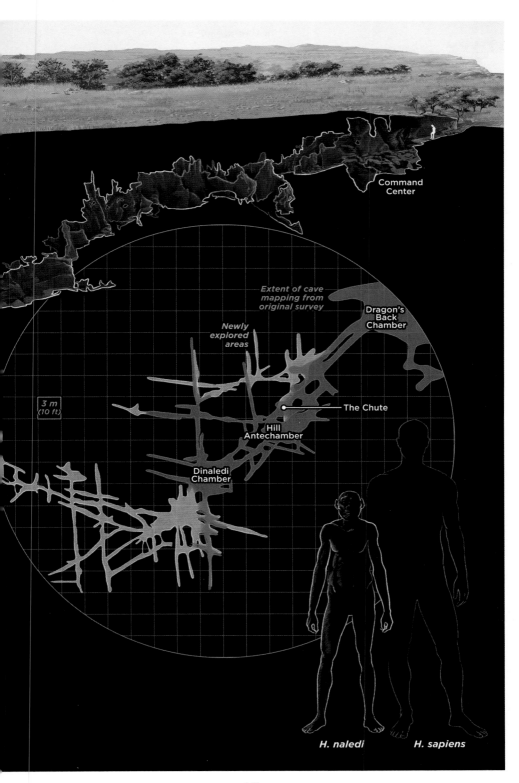

ABOVE: *The burial feature in the Dinaledi Chamber.* BELOW: *A 3D reconstruction of the same burial feature shows all the excavated bones in their original positions, including upper limb bones (left) and lower limb bones (center and right).*

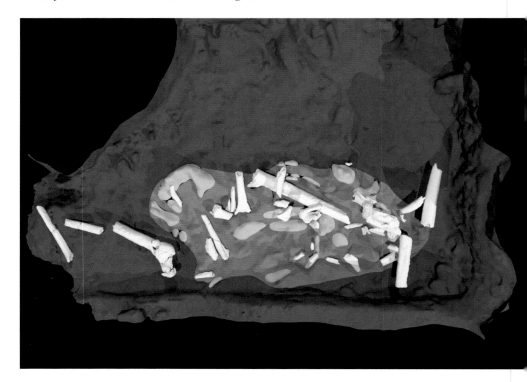

ABOVE: *Lee Berger examines the first engraved wall panel he discovered in the Dinaledi Chamber.*
BELOW LEFT: *The Dinaledi crosshatch (left), photographed with a polarizing filter. Toward the bottom, engraved lines bump into the fossil stromatolite.* BELOW RIGHT: *In blue, an outline of the Dinaledi crosshatch; in red, the comparable outline of the Gorham's Cave crosshatch, made by Neanderthals around 60,000 years ago. The similarity of these two cave markings surprised everyone on the team.*

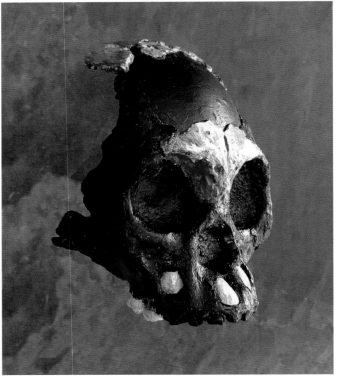

ABOVE: *A pile of rocks in the Rising Star Chamber shows signs of soot and burning—a possible hearth.*

LEFT: *A reconstruction of a skull found in deep fissures beyond the Dinaledi Chamber. Dubbed Letimela—"the lost one"—the skull's relationship to other finds is still a mystery.*

ABOVE: *The partial jawbone of a hominin, including at least two teeth, glows in ultraviolet light from the ceiling of the Dragon's Back Chamber.* BELOW: *Team member Sarah Johnson sweeps sediment from a possible fossil find in the Dragon's Back Chamber.*

Lee Berger, helmet fitted with a camera, prepares for the climb of his life, exiting the Dinaledi Chamber by climbing up and out through the Chute.

part of the chamber floor had been more than just gently settling dust. This was a disturbance. "So this is dug into that slope, right?" I asked John. "It's not a natural hole?" I was afraid to assume lightly.

"That slope is almost 45 degrees in Dinaledi," John said. "The sediment layers under the surface follow that angle. So how is this concave hole there? Besides, the angle of the foot doesn't follow the angle of the slope, either. It follows the bottom of the hole, cutting back into the slope from the opposite direction." He was right. This could not have been a natural depression in the slope. The remains visibly cut through surrounding layers of sediment that were parallel to the angle of the floor.

I was stunned. What he was showing me was the first clear evidence of remains that had not fallen down the Chute, come to rest on Dinaledi's surface, and then been covered by dirt. Somebody had dug a hole there, placed the body inside, and covered it up.

"This child is incredible," I breathed.

"*They* are incredible," John said.

My eyebrows lifted. "There's more than one?"

John began to rotate the image. He pointed out a second cluster of teeth near the perimeter of the body. "I think there are teeth from at least two more individuals in there. One is a child even younger than the main skeleton. Another might be older." I counted the teeth. He was right. I couldn't believe it. But John still wasn't done: "There might even be a fourth individual in here," John said. "I think I see tiny bones from what might be an extraordinarily young body. I've spent hours trying to confirm it, though, and I'm still not sure." He showed me some shadowy images. They might indeed have been the tiny tubular shafts of immature bones.

We spent the better part of the morning cycling through the images, flipping back and forth through the album like it was a yearbook from a 250,000-year-old high school. John continued to point out details as I tried to piece them together into a story. I saw a child's body, placed in a grave in something approximating the fetal position, with remains of two

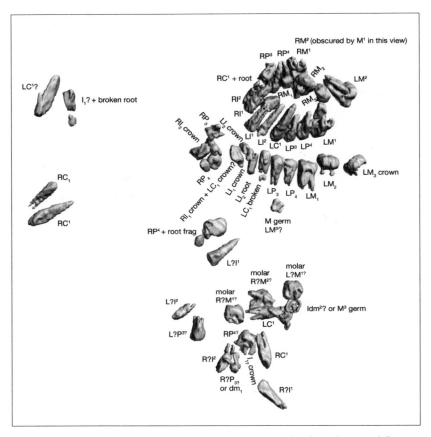

A reconstruction of teeth found in the Hill Antechamber burial, created from a medical CT scan, shows their orientation. They come from three individuals, and one set lines up as it would within a jaw and skull—part of the partial skeleton of a child.

or three others included in the same hole, or right next to it. As we moved from bone to bone inside the block, I tried to identify each one as part of the skeleton, but then John's gallery arrived at a crescent-shaped object, something denser than the bones and smack-dab in the middle of the most complete skeleton.

"What's that?" I asked.

"Ah." John broke into a knowing, satisfied smile and took a sip of coffee.

He had been saving this one for last. "That's a rock. Sitting right next to the skeleton's hand." He leaned back in his chair. "It looks like a tool."

I had no words to express my amazement. John's meticulous work on the CT scans had not only produced the most convincing picture yet of what might be a *naledi* burial; it had also revealed a possible stone tool within that burial. Our discoveries were growing more complicated—and more controversial.

| 10 |

A TURNING POINT

"The tools maketh the man." It's an old saying that was at the center of the belief of human uniqueness throughout much of the 19th and 20th centuries. It wasn't seriously challenged until Jane Goodall reported observing chimpanzees at Gombe Stream National Park in Tanzania creating termite fishing tools. Then the dam broke, and over the course of the next half century, study after study showed many animals modify objects into tools used for specific purposes. Even animals with small brains, like birds, we discovered, use tools.

But for some reason, the idea remained entrenched that tools were used only by the larger-brained species in the hominin fossil record. It wasn't until the last decade or so that it became clear that the use of stone and bone tools likely went back to some of the earliest australopithecines—hominins with brains no larger than those of chimpanzees. Yet the idea hung on that only the more human-looking hominins—those with bigger brains or smaller teeth—made more sophisticated tools. It permeated the belief system of the science. *Homo habilis*, not *Australopithecus boisei*, must have made the Oldowan industry—the name given to simple worked stones found in East Africa. *Boisei* simply didn't look like a toolmaker! We thought that *Homo erectus* must have made the Acheulean tools—more sophisticated hand axes—not because we had ever found clear evidence associating that species with the tools, but simply because among the species known to live during the period they represent, they had the largest brain. In short, in our field, increasing brain size was inextricably linked with growing sophistication in tools. It was another scholarly assumption potentially threatened by our discoveries—specifically the tool-shaped stone in the Hill Antechamber burial.

I used the data from the CT scan to 3D-print a replica of the possible tool. I wanted to hold something like what this *naledi* child might have held. In reality, the stone was about the size of a Swiss army knife—15 centimeters long, and four centimeters at its widest. It fit perfectly in my hand. It was thick at one end but tapered down to a point at the other, and one edge even seemed sharp, like a blade. It also had a characteristic divot in it. I wondered if that had happened during manufacturing.

If the rock was a tool, then it was the first artifact—the first *made* object—we had ever found in Rising Star with *naledi* remains. But was it actually a tool? When I showed my replica to experts in stone tools, some said it resembled a Middle Stone Age blade dated to perhaps 100,000 or 200,000 years ago. But when I told them where it came from, they backed away from calling it a tool. Some even urged me to leave the rock out of my review of the burials. Only the geological and skeletal evidence were relevant, they argued. The rock just confused the issue.

I couldn't accept that line of reasoning. The rock was in the plaster cast, mingled in with the bones. Anyone who looked at our scanning data would notice it. We had to include it in our description of the find, even if we still had many unanswered questions about it: Had it wound up there by accident, or was it some kind of offering—what paleoanthropologists call a "grave good"? Or had it been used to dig the hole itself? Still, it did seem to be positioned near the skeleton's hand. People in my field often joke that they never find a tool in the hand of the person who made it. Now that "person" was *naledi*.

Despite my self-assurance, I was prepared for an onslaught of skepticism once we published our findings about these *naledi* burials. *Naledi* was *Homo*, but it was far from human, with a brain one-third the size of ours. Scientists might accept that large-brained hominins like Neanderthals could exhibit complex behavior, but the idea that *naledi* engaged in anything of the sort was a harder pill to swallow. So many paleoanthropologists simply ignored the existence of this evidence.

As another saying goes, "The absence of evidence is not the same as evidence of absence." In the case of *naledi*, there is a difference between evidence that isn't good enough to prove complex behavior and proof that ancient hominins were never capable of complex behavior. Many archaeologists assume that complex behavior—deliberate burial, toolmaking, the use of fire, and the creation of symbols and art—was unlikely in early hominins, but I don't think it's out of the question. All these species were much closer relatives to us than chimps or bonobos, our closest living relatives, are today. And decades of studies of living primates have shown them to be remarkably sophisticated. Skeletal features of *naledi* allowed us to see it as a habitual biped with a brain shaped like a human brain, despite its smaller size. The small teeth of *naledi* would have functioned like our own, suggesting they were eating a humanlike diet. There's no question that *naledi* was much closer to a human than to a chimpanzee. Therefore, we thought, it is reasonable to think *naledi* behavior might have been complex. But what matters is what the evidence proves.

Everything we were finding in Rising Star was suggesting that *naledi* bodies were buried, but we needed to corral the evidence so that other scientists could test our conclusions, take a closer look at the possible tool, and share all our findings with new collaborators who would be open to our bolder theories. If we played our cards right, we could launch the next phase of our research to build a complete picture of this amazing species.

In 2022, we began assembling the data into a logical order, building our case on multiple lines of evidence. First, we organized all our information about the various possible burial sites. From the Hill Antechamber, we had 3D scans of the complete skeleton and its alcove, and from the deeper Dinaledi Chamber, where we had stopped excavating following my conversation with Kene, we had surface images of more than 50 collected fragments—some juvenile, some adult—suggesting at least two individuals.

The Hill Antechamber remains primarily consisted of a single articulated skeleton, but there were bones from other individuals in the space with it. Why? Perhaps several individuals had been interred together—as many as four children and possibly one fetus. And rather than thinking these individuals all died together, we thought it was more likely the hole had been dug into sediment that already included *naledi* remains. We've seen burial sites dug and re-dug throughout human history. Sometimes people choose to place new burials near old graves, and the bones of the deceased mix with the bones of ancient relatives. Perhaps the block housed a pattern of repeated use involving Hill.

This explanation might bring clarity to the mysteries surrounding the Dinaledi Puzzle Box, too. That area had contained articulated body parts of a few individuals. One leg of a child matched other bones nearby, but pieces from three adult skulls, a hand, a foot, and an infant's spine were in the mix, too. Maybe these weren't just a random assortment but were bodies that had been supported by soil before their soft tissue decayed and they collapsed into each other, disrupting what had been buried there before.

A TURNING POINT

The Dinaledi features we left partially excavated—the possible graves—also held a commingling of remains from different individuals. When parts of two or more bodies are jumbled together, certain bones are usually duplicated, but most of these remains didn't duplicate. The sediment near the surface held a few juvenile fragments that duplicated other bones in the feature, but that followed the pattern we were finding throughout Dinaledi, a possible repeated use of the same burial site for multiple individuals.

Kene led a team back to Dinaledi in March 2022 to inventory every bone fragment we had found before I'd called stop on the excavation. Her accounting provided another important clue: Toward the rear of the chamber, at the possible grave feature, she found a crushed cranial vault—the crown of a skull—as well as two articulated vertebrae and fragments of arm bones. At the other end of the feature, she recorded leg bones and other lower body parts. This configuration suggested that sediment had supported the body as it decomposed. The skeleton had collapsed downward, but we could still see some of the anatomical position that would have been present in a newly interred body. In other words, this position suggested an individual who was buried before its body began to rot.

When most people imagine a buried body, they think of a skeleton lounging with all its parts in textbook position, like in an Indiana Jones movie or *The Mummy*. But that's not how most ancient burials look. When ancient people buried a body, they usually made a much smaller hole than the stereotypical "six foot deep" that accommodates a coffin. The actual hole would be just large enough to hold a body crunched into a flexed or seated position, and deep enough to cover with dirt. As the bodies in this position decompose, the arrangement of the skeleton changes a lot. The soft tissue breaks down, the bones come apart at the joints, and gravity pulls them out of position. The dirt above the body begins to collapse into the spaces where organs and muscles used to be—sometimes carrying objects that may have been on the surface or in the dirt into the cavity. Over thousands of years, the bones themselves degrade into fragments as

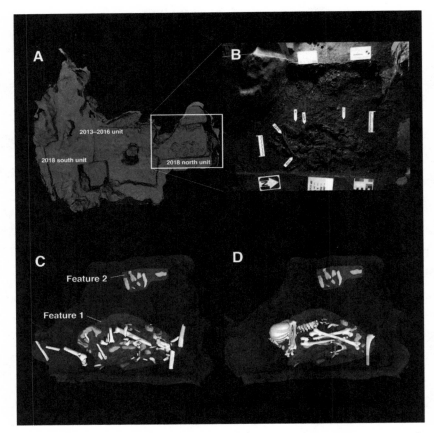

Four steps in studying one excavation site in Dinaledi: A: Work up to 2016, with an area of interest designated for future work. B: Further excavation in 2018 revealed an oval depression, likely a grave. C: An enhanced image shows the fossils found within the depression. D: An enhanced image of the individual (Feature 1), likely in a seated or fetal position. Fossils marked Feature 2 may represent another individual in a nearby grave.

the earth continues to compress them. With this in mind, what we found in Dinaledi were nearly textbook examples of ancient burial sites.

The sediment itself supported our theory. Geological study of the feature found a thin layer of clay under the undisturbed surface everywhere around the possible graves, but not above or inside the grave itself. Instead,

Kene found small chunks of this same clay *inside* the grave, mixed with the bones. Somebody would have had to disrupt that clay layer to blend it with the deeper sediment. In the broadest of terms, this was evidence that something had disrupted the ground here. It looked like we had a strong case for these being burial chambers. But we also had to deal with the tool-shaped rock.

In early 2022, I set up a trip to escort the plaster cast containing the rock to my scientist friends at ESRF in Grenoble. Their synchrotron, a super-powerful x-ray machine, can harness the radiation of überfast subatomic particles in order to—among many other things—look inside solid objects. It's spectacular science, but sometimes I prefer not to think about all that energy and chaos around my precious findings.

The first synchrotron scans gave us a problem, however. I thought the high-resolution scans would produce an ultimate view of the skeleton, and they did, but the scans showed the remains had shifted and compressed while inside the plaster jacket. It was going to be really hard to compare the new scans with the earlier medical CT scans John had showed me—we'd need to scan the entire block again using the old method. But those early synchrotron slices did confirm that the block contained very young infant bones, as John had guessed. They also gave us a high-resolution reconstruction of the tool-shaped rock.

Judging by the rock's surface, it was a piece of dolomite or chert—materials we were well familiar with after years of working in Rising Star. That meant the rock probably came from the cave system and wasn't carried into the cave from somewhere else—although we could not outright disprove that possibility.

One edge of the rock was indeed sharper than the other, like a blade, but the edge was irregular, suggesting wear. The scan confirmed the divot, or flake scar, that we had noticed before, but there were no other clearly

The rock found near a child's hand within the Hill Antechamber (top) looks similar to a stone tool found in Blombos Cave in coastal South Africa (bottom), dated to 78,000 years ago.

identifiable features that would prove the stone had been deliberately shaped into a tool. We asked expert colleagues of ours who were well versed in ancient African technologies, and while they agreed the object would have been useful and had characteristics consistent with a tool, its shape and makeup offered no definitive evidence of having been intentionally shaped.

In short, all we could say was that it was a tool-shaped rock, and it *might* have been used as a tool, and it was found near the hand of a *naledi* skeleton—in a grave.

We needed collaborators who could help us acquire a broader understanding of what *Homo naledi* burials would mean in the context of human evolution. We needed to talk to researchers who wouldn't shoehorn *naledi* into a given category. We wanted people who were willing to approach *naledi* on the species's own terms.

I reached out to Agustín Fuentes, an anthropologist at Princeton University who has spent most of his career studying wild primates, particularly how human minds might have evolved from these ancestral species. Agustín worked with a research team studying social evolution in humans and our extinct relatives. We invited him to collaborate with us, as well as Penny Spikins, an archaeologist from the U.K.'s University of York who focused on injuries or disabilities in ancient hominins, and additional researchers who specialized in burials of ancient hominins around the world.

We set up a meeting at Princeton in June 2022. Three of us—Agustín, John, and I—presented two papers, one on the technical aspects of the graves, and one on the broader implications of the discoveries. Everyone was intellectually supportive, but we wanted them to find the weak points in our arguments. Now was the time to hear feedback before releasing the papers for peer review. I ran through the lines of evidence that showed the *naledi* bodies were placed into holes and covered after their deaths. Agustín's team offered constructive comments about how we should present *naledi* behavior. They had a lot of helpful context from their research that connected what we were finding in Rising Star to the bigger picture of hominin behavior from other times and places.

As we presented the technical paper on burial, I kept an eye on Penny. I thought she would be one of the toughest critics of our work. She had investigated many Neanderthal burials, even studying the care given to the infirm before they died.

We finished the presentation, and I asked for comments and criticisms. For a long moment, there was a deafening silence. Then Penny put up her hand. A little reluctantly, I recognized her and gave her the floor, expecting to hear that she had caught a huge flaw in our data. Surprisingly, her message was the opposite.

"I think the data is good and convincing," she noted. What a relief.

But then she added a wry comment. "I don't think people are going to be surprised at the burials," she said, "but they are going to be wondering why there is no rock art on the cave walls." I laughed. Everyone in the room laughed. Burials were one thing, but art? Deliberate markings with intended meaning? Long considered a human hallmark? That was ludicrous.

The presentation went well, though a few inconsistencies emerged in our data that bothered me. Geologists suggested water had been involved in moving the bones, even though the sediment showed no signs that water had flowed strongly enough to move bones from somewhere else into the chamber. Others suggested a drainlike hole in the floor had sucked the bones downward, but we hadn't found any such feature during our excavations. Some still clung to the idea that the bodies might have been dropped into the Hill Antechamber and then carried into Dinaledi by the flow of sediment over the years, despite the burial data we were showing. But when I reassessed the map of Dinaledi and noted the two passageways that connected Hill with the deeper chamber, those features still seemed to create a choke point that would prevent sediment flow.

But we had been wrong about those tunnels already. Could I be sure?

The new theory exposed the limits of the data. Our ultimate goal was to tell a deep, engrossing, accurate story of who *naledi* were and what they were like—but in a few key areas, we were still guessing.

And there was one obvious way to eliminate most of that guesswork. As I again pored over the map, that familiar spark, that call to adventure, burst to life in my chest. My entire research team was preparing to go back to Rising Star, back into the Dinaledi Chamber, in just a few short weeks.

What I had not yet told anyone was that this time, despite all risks, I planned to join them.

PART III

JOURNEY INTO DARKNESS

| 11 |

TRAINING

The long preparation for my journey into the Dinaledi Chamber had really started in February, around the time I went to Grenoble for the synchrotron scanning. We had weighty questions to answer and a seemingly radical, potentially controversial claim guiding us: A nonhuman species with a brain barely larger than a chimpanzee's had buried its dead. This could change the story of human origins, and as a result, it could be examined and reexamined likely for as long as the archaeological community survived. The team and I had to make every effort to ensure we gave the world all available data in a clean, comprehensible way. I did not take the risks lightly, but I had to see all the evidence—which meant getting into Dinaledi myself.

Before I could think about navigating the perilous risks of such a plan, I had to worry about even fitting into the Chute. To be blunt, I needed to lose weight if I was going to have any chance of making it into Dinaledi.

The most versatile and adaptable cavers are like the inflatable men you sometimes see outside car lots—skinny and bendable, with perhaps an enthusiastic smile to match. By comparison, I looked more like the Michelin Man. Tight spaces like Superman's Crawl and Berger's Box gave me trouble, and to compound my difficulty, I was taller than most other cavers and scientists—not ideal for crawls where being small and compact is the greatest advantage. In fact, I would be the tallest person to ever attempt the Chute from my expedition team. But, I thought, if I could lose enough weight and add enough strength, I just might be able to fit. I was approaching my 57th birthday. I wouldn't have many years left to try. It was time.

Still, it was a tall order to get down to a size that might fit into Dinaledi, and I wouldn't even know whether I fit until I reached the critical halfway point of the Chute, where the squeeze was tightest. It would take enormous discipline to make these changes by the time our expedition began in July. For the time being, I kept my plan for attempting to get into Dinaledi private. Our work continued, and I told nobody—not the team, not my closest collaborators, not even my family.

As a researcher, I knew of only one way to approach my health: scientifically. I made my own diet and exercise plan that sorted out to one basic rule: Burn more calories than I consumed. This required a complete overhaul of my eating habits, and I switched my routine from three meals a day to four.

Each day began with a salad for breakfast—something that amused Jackie to no end—followed by an egg for a late morning snack. For lunch I turned to more protein to stay full, usually a tin of sardines or another egg, before I had an early dinner around 4:30 p.m. I ate earlier than normal to give my body more time to burn off the day's fuel before sleep slowed my metabolism, and I kept portions small and low-carb to keep my calorie count low. I occupied my stomach by drinking lots of water, perhaps eight to 10 liters a day.

Dieting can take you far in weight loss, and it also helped me to ramp up my exercise as I changed my eating. I work on a computer a lot, so to try to inject more activity into my day, I set up a small set of adjustable weights in my rooftop office, as well as a boxing bag and a set of resistance bands. Each evening, I would pause every 15 minutes or so to stand up from my desk and work out—just doing short, frequent bouts of exercise.

While my family gave me lots of cheek and lots of encouragement during my health kick, I still didn't tell them, or anyone else, about my plan to see Dinaledi in person. It was a pretty big secret to keep, but I had a few reasons for masking my true motive for weight loss. First, my wife is a medical doctor. She's seen many people, including our two children, Matthew and Megan, navigate the Chute. She's heard the stories of close calls and narrow escapes, and she even helped me formulate the safety protocols I used in the cave. I knew what she would have to say if she found out her 56-year-old husband was even considering the attempt, and it would be slightly more nuanced than unbridled support.

My kids were similarly aware of the dangers. Both Matthew and Megan had entered the Chute as skinny, fit teenagers, but even then they had talked about how hard it was. I knew this was going to be one of the most physically demanding things I had ever attempted, and likely one of the most mentally tough moments of my life. I didn't want anyone trying to talk me out of it or sowing doubts in my mind. I was already carrying a lot of questions about whether it was possible for me to make it down the Chute; I didn't want the slightest scrap of fuel for those reservations.

By late June, I had lost 20 kilograms—44 pounds—and I was feeling as fit as I had been for decades. My waist was down from a size 38 (a *tight* 38, I'd add) to a size 32, and my upper body strength rivaled what it had been when I was in the Naval ROTC. I was getting close to what I thought was fit enough, and skinny enough, to make the attempt.

Our first stop on the July expedition was the Dragon's Back Chamber, the space one enters on the way to the Dinaledi system. Our interest in the Dragon's Back Chamber had begun in March of that year. While Kene was inventorying the bones in the Dinaledi Chamber, John and I ventured into the Dragon's Back Chamber.

The ceiling of the Dragon's Back Chamber soars 15 meters above the cave floor, supported in part by a column near its center. On the left side of the chamber upon entering, you come upon breccia—the concrete-like rock that sometimes contains fossils—extending to the chamber wall, making a low overhanging ceiling. Near the Dragon's Back itself, the breccia slopes downward and forms a ledge, though a break interrupts the feature in the middle. Beneath the ledge, there's about a meter of space into which you can shimmy and look up at the breccia to see a layer of fossils. Many of these pieces might be carnivore, including a jaw with a sharp tooth poking out, but there's also a long, slender bone that looks like a *naledi* tibia. These fossils have tempted me for years, but in reality, there's little we can do to extract them. I once tried to free the fossils using an air scribe, a vibrating tool that causes rock to flake away from the fossil without harm, but when I lay under the ledge, I was overcome by the eerie fear that the block above my head could fall at any moment. So I nixed that plan. As a team, we pondered whether we could drop the block using light explosives or expanding material, but I felt hesitant about that. The architecture in these spaces is complicated. Destabilizing even one block might create a dangerous ripple effect across a greater area, and this was the only route in or out of the Dinaledi Subsystem. Risking a cutoff was unacceptable. The fossils stayed put.

Nevertheless, the bones in the breccia told a story about the Dragon's Back Chamber. These rocks were full of material from outside the cave system, with a totally different makeup from the clay-rich sediment in Dinaledi and the Hill Antechamber. That meant the sediment in the Dragon's Back hadn't flowed into Dinaledi since *naledi* had left bodies in the cave—making it hard for animals to enter, too. This was important for our burial-chamber hypothesis; it clearly established a permanent separation between the Dragon's Back and Dinaledi, and showed there had never been an easier entrance from this direction through which bones, *naledi,* or anything else could travel.

That also meant entry into Dinaledi had been challenging for *naledi* themselves. Based on our most recent dating of the rocks that formed below and above it, the Dragon's Back itself had likely fallen between 225,000 and 295,000 years ago—a range that overlapped with the age we estimated for the *Homo naledi* fossils in Dinaledi and the Hill Antechamber. We could not be sure whether *naledi* climbed the Dragon's Back, whether the block that formed this razor's edge ridge had fallen after the burials, or whether it had fallen sometime during the period when the graves were made. But regardless of the precise order of events, with or without the Dragon's Back, *naledi* would still have had to climb much of the way up the back wall and then down to reach Dinaledi. The entrance to those distant chambers had never been easy.

Understanding the ancient pathway into Dinaledi was important because for our trips into the lower caves, we hoped to find traces of the lives of *naledi,* not just their bodies. We knew that, like us, *naledi* experienced the Dragon's Back Chamber as the last stop before making the arduous climb into Dinaledi. Maybe they had used this chamber as a staging area, or a place where they ate and slept before beginning another leg of their journey. As I looked around the Dragon's Back with John and Kene and the others, I imagined strong and thin *naledi* individuals moving up the Dragon's Back block. They wouldn't have had to harness themselves in for the climb, but probably moved agilely upward using their curved fingers and toes. I imagined the flickering fires they must have built to light the way, small flames or torches set in the wall's little crevices or holes. The light would have created eerie flickering shadows of the climbers on the damp cave walls.

In the midst of my imaginings, as Kene and the others climbed toward the entrance to the Chute, John approached me. "Why aren't we digging here?" he asked.

I paused. He had a point. The Dragon's Back Chamber was a perfect place to excavate. We had been so focused on the precarious breccia slab that we hadn't thought about the floor of the chamber at all. Yet it seemed likely to house artifacts or other evidence of *naledi*. Why *hadn't* we been digging there?

The oversight was my fault. As a team leader, I had been more reactive than deliberate about our process. I had been responding to events as they occurred—the appearance of bones here, the emergence of a new theory there—rather than thinking critically about what areas *naledi* might have used as they traversed these spaces. Many of my early decisions had been guided by Steve and Rick's initial discoveries in the Dinaledi Chamber. They had seen dozens of bones on the cave floor—the same was true in the Lesedi Chamber—and we had simply become laser-focused on the challenge of Dinaledi and the ensuing questions it raised.

At John's question, I scuffed the floor with the toe of my boot. The burials in Dinaledi and Hill had been beneath five to 10 centimeters of dirt. What if there were graves just below our feet in the Dragon's Back Chamber, hidden only by this same layer of dust?

When I am looking at a rock that might contain a fossil or an artifact, I sometimes fool myself into thinking that I can see what is below the surface. It's as if the surface is the wrapper of a candy bar with a label telling you what's inside, and it boosts my eagerness to dig like nothing else can. Of course, rock and soil are actually impenetrable to the human gaze. It's why we use x-rays to look inside them. But the allure of untouched rock is enticing for paleoanthropologists, and one fantastic find can fuel a lifetime of obsessive searching. Once, my wife, Jackie, had scanned a rock no bigger than an American football to discover that a nearly complete and intact skull of *sediba* was hidden inside! We never would have known by looking at the surface. A millimeter of dirt or rock can hide something remarkable underneath.

Over the next several days, everyone bought into the idea that excavating the Dragon's Back Chamber was important for our understanding of how *naledi* used the system. We developed an excavation plan. Kene would lead the Dragon's Back excavation team in opening small squares of about two meters by two meters on the north side of the chamber. The work would be similar to what we'd done in Dinaledi—but we'd enjoy a much easier commute. Our teams would work for three weeks underground.

We weren't just looking for more bones, though. Anatomy and geology were the tools we had to make sense of what we were finding, but we wanted to apply a more behavioral lens to our work in the Dragon's Back Chamber, necessitated by what we were theorizing about *naledi* in other parts of Rising Star. I asked Agustín Fuentes to join our expedition, hoping he would prompt us to think in these fresh ways about *naledi*. Aside from his expertise, Agustín looks like he came straight from central casting for an explorer type. He's thin and wiry, with long, unkempt black hair. I thought his knowledge of the primate mind might contribute to a breakthrough, too.

Our July 2022 expedition would also host a group of documentary filmmakers. A production group had contacted me about telling the stories of *naledi* on film, and this seemed like the perfect opportunity for them to get some action shots of our fieldwork. Who knows? I thought. Maybe they'd be there when we made a discovery.

| 12 |

APPROACHING THE CHUTE

Expeditions are not simple to plan. There are accommodations to organize, research goals to coordinate, and crucial communication lines to build and maintain. We took months to prepare for a trip that would position the Dragon's Back excavation alongside another journey down the Chute into Dinaledi. The extra time bought me a cushion to try to lose more weight and prepare for more caving. One day in late July, the entire team assembled on-site ready to get to work. I suited up in my jumpsuit and accompanied the team underground for the opening of the excavation. It was a day full of systems checks and familiarizing new team members with one another and the cave. Everyone was excited, but the experienced researchers among us realized it might be a while before anything new turned up.

On the second day, Agustín, John, and I geared up and went into the Dragon's Back Chamber. As the rest of the team worked on the floor of the chamber, we put on harnesses and prepared to climb the Dragon's Back formation itself. I intended to guide John and Agustín up the ridge to help them take in the excavation from a different vantage point, but I also had a secret agenda for this trip: I wanted to test my body in the top of the Chute.

The three of us spent an hour at the top of the Dragon's Back, discussing the area and looking at the various fissures below us. Agustín looked for a long time down the narrow crack that formed the entrance of the Chute, contemplating how difficult it must have been for *naledi* to access the system of chambers below.

Once we had finished our debrief, I let the others head back down ahead of me. But before I brought up the rear, I turned and slipped into the tight confines of the crawlway, slithering on my back in the tight space. I maneuvered over the open fissures and felt with my boots the Chute opening. I let my body slide down until my chest met the gap. It was tight. The rock touched both the front of my chest and my back. I leaned over and peered down into the fissure below, my feet finding tiny protrusions on the rock wall. My headlamp lit a turn and a tight squeeze about four meters below me. Beyond that, somewhere out of sight, I knew the dreaded 19-centimeter squeeze awaited. Like someone memorizing a sequence of dance moves, I went through in my mind the motions I would do to navigate the passage.

I had been telling the world how dangerous this passage is since day one of the first Rising Star expedition. I had been the one to write the safety plans in the event that someone became too injured in the chamber to get themselves out. Extraction was not an option; it wasn't possible to assist someone in the tight spaces of the Chute, so hurt cavers would have to either make the climb on their own or stay down in Dinaledi. The rescue plan, if you could call it that, would be to send a doctor down into the subsystem to stay with the injured party until they could climb out under their own power. Fewer than 50 people had ever been inside Dinaledi, and I would be not only the largest but also among the oldest.

I wiggled from side to side in the tight confines. I pulled myself up, then lowered myself again. My body felt strong. I was as ready as I'd ever be.

APPROACHING THE CHUTE

Looking up from the Dragon's Back Chamber, one can see the ridge that gave this formation its name (left of center), along which excavators climb to reach the Chute, which leads to Dinaledi. Wires for gear transport and communication stretch through the passage.

The next morning the whole group assembled in front of the Exploration Center, our permanent base camp just outside the cave entrance, for the daily briefing. "I'm going to make an attempt to get into the Dinaledi Chamber tomorrow," I announced. I saw the glances among the exploration team. Kene smiled uncertainly. She had been through the Chute many times, and even after my weight loss, I was probably twice her size. Maropeng—perhaps the most vocally enthusiastic team member—smiled as he said, "We'll get you in, Prof." But his face betrayed his disbelief. John and Agustín nodded—I had told them of my plans the night before at dinner. Dirk van Rooyen, the team leader that day, would be in charge of my descent and ascent, and I had consulted with him the previous afternoon to make sure he thought I was fit enough to make the journey. He looked me over. Whether he thought I was a fool or not, he didn't say, but his professional judgment deemed I was physically ready to try. Most important, he said that I wouldn't be putting the team at risk by doing so.

The person who looked the most shocked was Warren Smart, the cameraman with the production team. He was an experienced adventurer and diver, and the producers had brought him on the project specifically because of the extreme caving involved. He had undertaken the Chute descent earlier in the spring to film Kene's work in Dinaledi, and afterward he spoke about the experience as an almost transcendent moment. He also swore he would not make the attempt again. It had pushed his physical and mental limits to the edge of safety.

But after my announcement, Warren approached me. "Hey," he said. "If you're going in, I damned well am going in, too."

I blanched. "Don't take any risks on account of me," I replied.

"It's my decision," he said. "No one urged me, no one bribed me, no one even tried to convince me. In fact, your buddy John already assured me I didn't have to go."

APPROACHING THE CHUTE

Agustín Fuentes

"You really don't," I said.

"I know." He smiled. "But I'm not missing the chance to take a camera down there with you."

John Hawks

On the day of my attempt, I woke up at five in the morning and began getting ready. I pulled on the long hiking skinsuit I wore while caving, then my jumpsuit. I spent half an hour checking batteries for my helmet light

APPROACHING THE CHUTE

and the other gear I would be taking in my backpack. I laid out the rest of my equipment and checked each device before packing it: a small humidity and temperature monitor, a powerful UV flashlight, two smartphones (wrapped in bubble packaging) to use as recording devices, a fully charged power bank, charging cords, a lightweight spotlight, my trusty Mickey Mouse watch, and a water bottle.

I sat down on the bed to lace up my calf-high British Army boots, given to me by a friend more than a decade ago. I had made hundreds of trips underground in them. Once I was done, I sat on the bed and looked about the room. I still had half an hour before I was supposed to meet John and Agustín for breakfast, so I stared at the walls of the B&B and tried to think positive thoughts.

My mind went to Jackie, probably just waking up to head to work at the hospital. I thought about Megan and Matthew. Megan would be in class in Gatton, Australia, where she was enrolled as a veterinary student, and Matt would likely be finishing up a production day on his latest movie—he was in a film program at the University of Southern California. I still hadn't told any of them what I was about to do. We are an extremely close family, but I knew they would raise their eyebrows at the prospect of me making this attempt. I had enough doubts about my ability to get through the Chute. A small part of me knew that it probably wouldn't take much for a close family member to talk me out of it.

For more than eight years I had been telling the world I would never get into Dinaledi. I had said that directly into the cameras recording the original expedition. And I had joked at almost every public lecture that my ego alone was too big to fit through the Chute, much less my body. In a way, it was how I treated my own disappointment with myself. The Dinaledi Chamber had changed my life, but I thought I would never go down there. Until now.

At breakfast, Agustín paused in the midst of drinking his coffee and gave me a pointed look. "You going through with this?" he asked.

I nodded. "I'm going to at least give it a try."

He nodded back. "OK." His smile seemed nervous.

It was a misty morning on the African Highveld. When we reached the dirt parking area at Rising Star, I stepped out of the jeep and stretched, watching the antelope on the property heading down to the watering hole. It was a beautiful, peaceful scene. John and Agustín headed over to the Exploration Center to chat with the film crew and the other explorers. I stayed behind and began preparing my kit, double-checking lights and helmets and batteries and phones, and arranging everything inside my orange backpack. When I was finished, I saw that Warren was alone at the production van, checking his camera equipment. I walked over and interrupted him.

"Hey, brother!" he said cheerfully. His one-piece overalls hung from his waist. His easygoing manner made me smile, but I wanted to offer him another out.

"You know, if you don't feel like doing this, I'll tell the director I have forbidden you from going in for safety reasons," I said. "He'll have to accept that. I won't give him a choice."

His smile only grew. "No way, brother." He placed his hand on my shoulder. "While I am grateful for your concern, I wouldn't miss this for the world."

I nodded, reassured that we had given him every chance to back out with dignity. I returned to checking my equipment and getting my mind in the right space.

I called everyone to assemble in front of the Exploration Center for my usual morning briefing. I used a whiteboard for the daily plan, and on that day, it included something I had never written before: my name as a part of the Dinaledi Chamber team. "I'm going in," I repeated, letting everyone know I hadn't changed my mind. "Like always, it's safety first. Dirk is leading the Dinaledi team. Kene and the Dragon's Back

Chamber team are all set." I handled the other formalities on the morning plan, including discussing who would be "topside," manning the Command Center, and who would be the runner through the cave, bringing things back and forth. I reiterated the safety protocols. I capped the whiteboard marker.

"See you in a few hours. I hope."

| 13 |

INTO THE CHUTE

Most sounds don't travel in caves. As I came out of Superman's Crawl and stood up in the Dragon's Back Chamber, I knew the whole film crew and excavation team were around the corner ahead of me. But the only sign of them was a soft glow of light reflecting off the high, vaulted ceiling.

I walked over to the right, where I could look down across the lower north branch of the chamber. This was the excavation area, where Kene's team was already at work. Her three team members each had a half square to themselves, and they crouched over them with trowels and brushes in hand. John moved about on the edge of a pit, taking photographs, while Kene took down records in her field notebook. Inches above the surface of the cave floor in this part of the chamber, the heavy yellow monofilament line that made up the excavation grid stretched into neat squares.

I ducked under the low overhanging ceiling into the south branch of the chamber. This side of the chamber was dark, lit only by the residual glow coming from the far end of the excavation zone. We used this part of the chamber as a staging area, full of equipment and space where the excavators could prepare for the day. This morning, it was crowded as never before. The production crew was recording the exploration team as they geared up for the day's journey into Dinaledi. Headlamps flashed as Maropeng and Warren pulled bright red climbing harnesses over their coveralls. Dirk was already halfway up the Dragon's Back formation, preparing to haul up equipment with the winch system.

I picked up the harness I would wear to clip into the safety line we had installed on the Dragon's Back several years before. The harness would be the only thing stopping a serious, possibly fatal fall if I slipped on the way up. I put my legs through the straps and brought the harness to my waist, clipping the belt in front and pulling it tight. John appeared from the excavation side of the chamber and approached me. He ushered me away from the cameras. "I think you're going to want to see this," he said.

I followed him to where Kene's team was working under the lights. John stepped over a fallen dolomite slab that had, at some time long ago, dropped from the cave ceiling and now marked the edge of the excavation area. Kene stood near the edge of this slab. "We are hitting some bone," she said. "And you might want to take a look."

I knelt down next to her. The bones laid within a discolored patch of sediment. The excavator had created a neat, flat surface around 15 centimeters below the floor level, sprinkled with small pieces of bone. Around the pieces, the soil's typical red-orange color was tinged with gray—it was a potential sign of ash. A small flake of bone had broken away from the larger fragment embedded in the dirt. The staining on the flake had discoloring that resembled the telltale black-blue we see on bones that have been exposed to fire.

"It *could* be burned," I said carefully. There was plenty of reason to be cautious. Sometimes minerals can stain a bone, and I didn't think we could possibly be lucky enough to hit something as significant as genuine burnt bone so quickly in the excavation. In general, proving the past existence of controlled fire is a topic of contention for paleoanthropologists. It's even more contentious in the context of cave work than in work on open land.

INTO THE CHUTE

Open landscapes introduce a huge number of variables that confound the explanation for how and when an ancient fire began. Fires in open or exposed spaces could have originated from burning tree trunks, wildfires, or even spontaneous explosions from deposits of bat guano—yes, it's true: When bats defecate in caves with poor ventilation, the buildup of methane gas eventually causes the material to self-combust.

Caves eliminate many of these variables—except the bat guano, unfortunately—but prehistoric cave fires are highly unlikely to leave a trace. These blazes would be small, meant to use just once for lighting a torch or cooking a meal. And since compelling evidence for fire, like the accumulation of soot on a wall or ceiling, theoretically takes many repeated fires to appear, it's always easy to eliminate fire as a possible cause when we find such discoloration.

But although there wasn't enough substance in Kene's finding to draw a conclusion, I didn't want to dismiss it outright, either. *Naledi* used multiple parts of the cave system where natural light didn't reach, an almost certain indication they had fire. But overall, evidence of the use of fire by specific hominin species is rare. Researchers suspect fire was used by earlier hominin species, such as *Homo erectus,* but it remains unproved, with little evidence associating *erectus* bones directly with fire. The situation in Rising Star pushed us to a logical conclusion: Any hominins that used the dark areas of a cave like this must have been able to light their way, and the only hominin we knew who had been this deep here (besides us modern cavers) was *Homo naledi.*

I dusted off my hands and stood. "Thanks for the heads-up on this," I said to Kene. "I'll take a closer look when I come back—*if* I come back."

She smiled. "I'll see you soon, Prof."

I walked back toward Warren. "Ready?" I asked. He had finished getting into his harness and was shouldering his camera gear.

"You bet!" he said. His face held zero signs of worry.

"OK, then let's do this," I said. I grabbed the fourth rung of the aluminum ladder and hooked myself into the static line to climb the Dragon's Back.

You move along the static line as you climb by holding on to a pair of metal clips connecting the line to your harness. You slide the clips to a stake where the static line is fixed to the rock, unfasten the clips from your harness, then refasten them on the other side of the stake to continue climbing. But there's one rule: Never unfasten both clips from your harness at the same time. You keep a strange rhythm: climb, unclip, clip, unclip, clip, climb. As I journeyed up the Dragon's Back, the deliberate movements gave me time to look at the falloff on both sides and think about *naledi* climbing this sharp rock wall in the dark. Did they scramble up the wall like gymnasts over the surfaces, unafraid of the heights or the potential for a fall? Or, like me, did they move carefully and deliberately, always conscious of how one mistake could deliver a brutal injury, or worse?

At the top of the climb, I pulled myself onto the last pinnacle and came to a one-meter bridge. This short walkway spanned a chasm plunging to the floor below. We used to take a leap over this gap, but I was trying to take fewer death-defying jumps on this side of 50, so I had had a bridge put in. Now the crossing was much safer, and each explorer could hook themselves on to more nearby static lines just in case.

As I crossed the bridge, Dirk, Maropeng, and the rest of the cavers stepped out of their harnesses and moved toward the Chute entrance. This was the last open space we'd be seeing for a long time. From here on, the journey would be through ever narrowing cracks and passages, with small openings giving glimpses into the network of faults and fissures below. I removed my harness as Warren approached with his camera.

"You want to go in front or behind me?" I asked.

"It would be great if you let me go ahead," he answered. "Then I could get your first reactions when you enter the chamber."

INTO THE CHUTE

"Don't count on that happening," I joked. I pointed him into the fissure that headed toward the Chute entrance, where Dirk and the others were preparing to descend. "After you."

The Chute entrance is a small opening one meter long and maybe 30 centimeters wide. The best way to enter is to wriggle in on your back, feet first, pushing yourself along with your hands against the walls. To your right as you slide forward, there's a crack, about 15 centimeters wide, that you have to scoot around to avoid being wedged into it. At the end of this short section is a small nook beside the Chute entrance where the Chute Troll crouches and watches your back. Maropeng took that position for our attempt.

Dirk slid into the tunnel after Maropeng, pushing a small dry bag containing our equipment: lights, cameras, and batteries, among other gear. The explorers had moved much of the heavier gear into the chamber the day before, setting up the internet system and testing that power was working. Equipment transfer would take some time, so I crouched and tried to get comfortable on the rock as I waited my turn.

Dirk went down, then team member Mathabela Tsikoane, who was a good foot shorter than me and probably half my weight. He struggled to make the first tight turn. I heard him grunting as he pushed himself past the entrance of the Chute. It made my stomach churn to watch him struggle. A seed of doubt sprouted in my head. Could I make it? Yesterday, I had slipped into that space with confidence. Now, perhaps not so much.

Mathabela made it through the Chute after 10 minutes of contorting himself through its maze of hairpin turns. Then it was Warren's turn. I nodded to him. "Be safe," I said. He gave me a thumbs-up, then squirmed through the opening until his light and helmet disappeared.

Doubt crept across my face. I looked at Maropeng. He caught my eye and said, "You got this, Prof." I tugged the sleeves of my jumpsuit. A few minutes later the intercom announced Warren's arrival in the chamber. I took a deep breath, rolled onto my back, and inched my way into the Chute passage.

| 14 |

THE DESCENT

"Your lower legs are in the Chute," Maropeng told me. "When you lower yourself in, there are some knobs of rock that you will feel. Use those to guide your feet down."

I nodded as my pelvis and stomach eased into the tight space. The rough rock scraped my stomach and back as I turned to face Maropeng. "Like this?" I asked.

Maropeng showed me on how to twist my boots to fit into the slot at the top of the Chute. I turned my ankles and began to bend my legs to a 90-degree angle. My gloved hands scraped the dusty surface, and the odd angle of entry forced me to shove my face into the rock. I tasted dirt. I slid downward until my chest caught on both sides of the Chute entrance. I twisted and pushed until the dark wall of the tunnel overcame my vision. I was in.

I had spent hours looking down into this space from above. I recognized the small knobs of rock that offered me toeholds and handholds, but I hadn't expected the walls to be so damp. I struggled to find purchase on the slick surface, closing my eyes and feeling for a protrusion that would grant me traction. I found a foothold, then lowered my whole weight onto it. My hands braced against both sides of the wall.

"At the bottom, Prof, you are going to see a gap that is behind you and to your right," Maropeng said.

There was hardly space to move my helmeted head. I shone my light toward my feet and saw an alarmingly small slot in the rock. "OK," I replied. It was the only way down.

"You need to get your boots into that, then rotate so you are facing the other wall." Maropeng's headlight cast odd shadows from above.

Maropeng guided me for the first few meters, helping me find footholds and handholds and body positions. Once I reached the first turn in the descent, about three or four meters down, I slid to the left. This was the first real squeeze, and if I got through it, I would disappear from Maropeng's sight. Dirk sat about four meters below that point to guide my feet and body through the next portion of the descent. After that, the infamous 19-centimeter squeeze waited, the most significant challenge of the entire journey.

I've been in a lot of dangerous situations during my 33 years as an explorer. Danger tends to develop fast, like when you turn a corner and encounter a ferocious animal. Our sort of exploration has risks, and cavers train for them. We practice ropework on trees, we build artificial replicas of the squeezes we'll encounter out of timber and plywood, and we rehearse practice scenarios of the worst-case circumstances. One of the original "underground astronauts" at Rising Star, a caver named Hannah Morris, practiced the Chute squeeze by wriggling beneath the chairs at her kitchen table. Regardless of how much you practice, though, little can prepare you for the real thing.

With difficulty, I lowered myself until my boots could twist into the gap. They barely fit, even when I pressed them together. I could hear Dirk moving below me in the darkness. "How's it going?" he shouted.

"So far, so good!" I yelled. I grunted again as I slid my legs into the gap. The rock pressed against my thighs—it was going to be an even tighter squeeze for my chest. By this point, my feet had cleared the gap and were dangling in space—no walls to touch or gain a toehold. I was about to make a major commitment. If I continued to lower myself, I would reach a point where this space compressed my chest and I would have no choice but to shove the widest part of my skeleton through the slot. I grimaced. This route would be my exit, too, and I knew that getting back up would be harder than going down.

Dirk offered words of encouragement from below. "OK, about half a meter below you is the top of a pinnacle—it's the sharp tip of a stalagmite," he said. "You will be able to get a toe on it. When you do, you want to turn around so that when you descend to your left, you are going to be hugging the pinnacle once you are in the space." I took a deep breath and envisioned the contortions he was describing, imagining the rounded tip of the lime cone that must be protruding just below me, unseen in the darkness.

I closed my eyes, then started to wriggle my way into the gap, my right toe stretching for the tip of the pinnacle stalagmite. I found it as my chest entered the tight squeeze. I pushed myself down with my arms, and with each breath my weight compressed my chest smaller and smaller. I felt the rock scrape against my back and sternum, pushing the bones inward, caving my skeleton. With the very tip of my boot, I felt the top of the pinnacle. "Is my foot on it?" I asked Dirk.

"Yes, that's it. Now what you need to do is rotate to your right so that when you drop into the space, you can hug the pinnacle with your back to the wall."

I grunted. Those moves would be worthy of a contortionist. With great difficulty and an intense strain on my torso, I managed to follow his instructions. At every point I kept my toehold, like a slow-motion ballerina corkscrewing into the rock. I inched my body into the position Dirk wanted.

"That's it," he said. His voice was my only reference. I could barely turn my head in this position. "Now lower yourself backward and to the right," he continued.

I took a shallow breath and pushed my way into the space. This was nuts. My boots felt both sides of the fissure. I knew in a few minutes I would have to get my pelvis and upper body, and then my helmet, through the 19-centimeter squeeze. This was the real point of no return.

I found myself in a space that immediately surprised me. I could finally see this peculiar stalagmite I had heard so much about. I could never quite picture the odd position it was sitting in, nor how close I would be to it when descending. I was literally hugging it. My legs angled down into a tight space and my pelvis was wedged into the narrow gap where my boots had scraped before. My cheek was actually pressed against the wet rock. As I caught my breath, resting, I looked about. This space wasn't a chute at all, I realized. It was nothing like what our geologists had described, and it was even different from the drawings of the Chute in our scientific papers and articles.

Steve and Rick had first discovered the Chute on September 13, 2013, and ever since then, based on their initial account and those of all the cavers who had followed them, we had described the Chute in every single paper as a chimney, a single vertical passage. This image became ubiquitous—even Wikipedia described the Chute in this manner.

But it wasn't a chimney. The Chute was an intricate network of potential passages. To my right, my light shone upon an extensive fissure that undulated with varied gaps and squeezes. Up above me, I saw other possible entrances into that network, and some of the openings even seemed large enough for *naledi*-size individuals to move through. I envisioned *naledi* scrambling around me through these spaces—adults and children climbing through whichever passage they preferred. They didn't have to go one at a time as we relatively bulky humans needed to do. They could go together, in parallel. It was a labyrinth of opportunity. I began to second-guess, as I hung there against the pinnacle, whether the Chute we had spent years cramming ourselves into was even the easiest way.

Armies are hard to deter once they start moving in one direction. The scouts that find paths don't always find the best route, yet once they blaze the trail, the generals are unlikely to seek out a better alternative. This is a phenomenon I think of as "pathfinder syndrome," a kind of inertia that explorers must watch out for if they want to stay clearheaded about their work. Our team followed the Chute because they had learned it from those who had gone before them. In fact, early in our explorations, we had scouted a second way into Dinaledi through the Rising Star system for our power and internet lines. And although that pathway was less passable than the Chute—only the skinniest cavers could fit through it—and was still viable, we still dismissed it all the same. But hanging in the Chute, I saw how foolish that had been. These routes weren't separate chimneys or tunnels; they were just two points within a vast and complicated network that only became constraining when we decided, arbitrarily, that we needed the widest paths. But *Homo naledi* likely didn't share our selectiveness.

Another thing I saw, hanging there in the dark, was that it wasn't a straight drop into the chamber below. I wasn't yet in Dinaledi, but I could already tell I had moved several meters laterally during my descent. Something dropped into the top of the Chute wouldn't be guaranteed to fall toward the Hill Antechamber below. Gravity might even pull it away from that direction, jamming it into a fissure on the way down or diverting it in another fashion. One thing I was sure of: The name Chute needed to go. It had been misleading everyone.

I continued downward. My hips passed through the 19-centimeter squeeze, but as I slid my chest into the gap after them, a cruel knob of rock jabbed into my sternum. I felt the bone bend. "This rock knob won't let me pass!" I cried. I could see Dirk's light flashing as he looked up.

"I know the one," he answered.

"I'm not sure I can get by it!" I said. The protrusion bit into my sternum. I struggled to pull myself back up, thinking I could try another way. I moved sideways.

"Don't go that way!" Dirk warned. "You're going to get into a space you might not get out of."

I slid back, panting, and contemplated my options. Looking up, I saw a half silhouette of Maropeng sitting above me, near the entrance to the passage. To one side of me was a climbing rope used to guide dry bags through the caves. It stretched from me to Maropeng. I had room to grab the rope with only one hand, my right, so I reached across the confines and wound the rope around my wrist as much as I could. I looked up. "Maropeng," I called. "When I tell you, can you give me a pull? I am trying to free myself from a knob of rock!"

I felt the rope tighten around my wrist. "Sure," Maropeng answered.

"Pull!" I shouted. The rope went taut, and I pushed hard with whatever leverage I could find with my legs and my free arm. It was just enough effort to lift me an inch or two, freeing my chest from the knob's bite. I wedged my hips into place just below the pinnacle underneath me, my legs splayed to either side of the passage. My shoulder twinged with pain from the rope's pull.

I looked at the impassable knob, my mind racing. I had assumed for nine years that the Chute was a special pathway. I had thought many times of *naledi* possibly moving along this path, maybe even carrying a body into the chambers at the bottom. With that in mind, I had long considered this place as somewhere important to preserve so that we could understand *naledi* behavior. I had long decreed that we could not change the route to make it easier, or more passable. But my new insights showed me I had been wrong about this passage's uniqueness. There was nothing special about it in the context of our work except that it could fit human beings. It was just one optional route amid a vast web of routes—nothing sacred. There was no need to preserve every detail of its rock surface. We had been making this journey unnecessarily hard on ourselves. I made a decision. "Dirk, can you take this knob off?" I asked.

If Dirk had any reservations about damaging the passage, he didn't show them. "Sure!" he said easily. "Give me a moment." He scrambled down into Dinaledi and retrieved a rock hammer. Then, with nothing more than a few swift strokes, he broke off the pestering chunk of rock. "That should be enough," he said.

After Dirk slipped back down to his previous position, I took a deep breath and again slid toward where Dirk had removed the knob. The broken edges tugged at my jumpsuit—but this time, though the protrusion still pressed on my chest, it no longer caught my sternum. I pushed all the air out of my lungs and sucked in my chest as far as it would go, at the same time summoning my strength to push myself downward. The rock scraped at my body. I clenched my jaw with pain. But then I was free.

The team had told me that would be the worst squeeze of the journey, and I was through it. I continued picking my way downward, my body seemingly contorting and compressing itself through each gap like toothpaste in a gnarled tube. Dirk fed me instructions from below—move a couple meters to my left, turn around 180 degrees, feel for a toehold—and after a few more minutes, the tip of my boot brushed the top of a ladder.

I could hardly believe it. This was the ladder that our team had specifically designed to help in this space. We had brought it here in 2013. Whenever a team member made it this far, our team issued a call from the Command Center through the intercoms across the expedition. I had heard it hundreds of times: "Marina has reached the ladder," "Becca has reached the ladder," "Kene has reached the ladder." It was the introduction for every explorer into Dinaledi, a signal that they had made it down safely through the Chute and that their grueling passage was over. Against all my instincts, I relaxed and slid downward, and then, all of a sudden, my feet found the top rung.

"Berger has reached the ladder."

I stepped onto the floor of the chamber and closed my eyes. Tears welled. For more than eight years—ever since its discovery—I had believed that I would never set foot in this space. The journey had been horrible, but I had already learned so much. The pain and fear were already worth it. Now I needed to make the most of the exploration period ahead.

More than ever, I was thankful we had established a Wi-Fi link from inside the cave. I pulled out my phone and dialed my wife's number on a video call. Jackie answered, and I smiled at her, my face filthy and sweaty, my voice full of elation.

"Guess where I am," I said.

"In a cave?" she quipped.

"I'm in the Dinaledi Chamber," I said. "I got in!"

Surprise shot across her face. Her words caught. Then she gathered herself. "And getting out?" she asked.

She was ever the pragmatist. I smiled. "If I can get in, I can get out," I said. I promised her that the moment I was back up the Chute, I would call her to confirm that I was safe. After a few minutes of chatting, we hung up. Now I needed to explore.

| 15 |

EXPLORING THE HILL ANTECHAMBER

Before my descent, I vowed not to take pictures or video of Dinaledi. I knew that this would likely be my only journey into the space, and I didn't want it filtered by a lens or a screen. A digital wall had stood between me and the reality of these caves for more than eight years. Now I was aching to just walk around and explore with my senses. The only step I took for posterity was to use the voice recorder on my phone to narrate my observations as I explored Dinaledi. I wasn't sure whether the result would be useful in the end, but I hoped it would heighten my powers of observation and help me stay in the moment.

I took out my phone and a laser rangefinder and began to dictate a description of the Hill Antechamber. It was smaller than I imagined: just a meter and a half wide and three to four meters long. I noted small side passages that seemed to be blind alleys. One of these stretched off to my right for five or six meters before disappearing into the distance. The passage itself was narrow, with a dropped block of dolomite in it.

It seemed Hill was at the center of a network of tiny passages. Seeing these openings, I reflected on my experience descending the Chute. *Naledi* was smaller than the average human in body size, with a head not much bigger than a softball. I could choose only one path to enter this space, down the most direct drop. But through the crisscrossing warren of fissures above me, *naledi* might have had other options. Perhaps there were many ways for them to navigate these spaces.

I looked up at the ceiling, a roughly triangular peak that stretched four meters above where I stood. Spectacular stalactites hung down like antlers, their tips clustered with bright white calcite. The chamber walls were made of bare gray dolomite—on the far side, bands of dark, quartzlike chert ran through the wall toward the Dinaledi Chamber. In the floor before me was the meter-wide rectangular hole where we had excavated the *naledi* child's skeleton and the possible tool. The bottom of the excavation was a perfect horizontal surface cut into the tilting floor, emphasizing the steepness of the 45-degree slope.

But the slope wasn't going the direction I expected. Our scientific papers had described the high point of the slope as the bottom of the Chute. But I could now see that the dirt came out of a different entrance, one with no link to the Chute at all, and likely no link to the Dragon's Back Chamber, either. This contradicted one of the leading theories about how the bones entered the depths of the Dinaledi Chamber. In the face of the burial theory, many skeptics had posed that *naledi* didn't climb down the Chute at all. Rather, they hypothesized that *naledi* tossed bodies down the Chute, whereupon gravity piled those remains into a heap and sent them flowing throughout the space. But seeing Hill myself, I could plainly observe that debris couldn't have fallen in the way those skeptics described. There was no slope going down from the foot of the Chute. I was beginning to understand the Chute not as a pipe that could carry material down into Hill, but as part of a network of many faults and passages. There was no chance it was a disposal pathway for the tossing of bodies into Dinaledi. The dirt and debris that had percolated into Hill over millennia could have come from all parts of that labyrinthine system.

I turned around to look at the back wall of Hill, the side where the Chute exited. Dirk and Warren sat on a broad shelf of flowstone that hung a couple of inches above the chamber floor. This shelf was less than an inch thick but obviously very strong—it could support the weight of several cavers. We had excavated the dirt on top of it during our 2017 fieldwork to check for artifacts, but had found none. Looking at it now, however, for the first time with my naked eye, I noticed the flowstone appeared broken away in places. One spot seemed like a fresh break, but the rest seemed to have been chopped away.

I took a closer look at the shelf's edges. The flowstone didn't end in a naturally worn way. I would have expected a curved, eroded surface, but this edge was sharp and jagged. It didn't seem like a natural fracture—someone or something had broken the flowstone here. But that raised the question: Where were the broken pieces? We had never found anything like that in Hill before; the pieces were missing.

Flowstones don't just break off, certainly not stones strong enough to hold grown men. A falling rock might have been able to do it, but I didn't see any obvious possibilities for that, and besides, the missing plate should be lying right there. I probed the explorers: Could this have been our team? Nobody took credit for the break. I wasn't surprised. The edges were obviously not fresh breaks—instead of showing clean white lime, they were covered in dirt, likely thousands of years' worth of dirt. If it wasn't us, I had to consider another possible culprit: *Homo naledi*.

To me, it made perfect sense. To reach the soft dirt on the chamber floor, *naledi* had to break through the overlying flowstone, probably using a large stone as a hammer. But that didn't explain why the missing pieces were gone. Could *naledi* have moved them somewhere else in the system? Or were the pieces in fact buried within the sediment? Solving this mystery had massive implications for the dating of *naledi*. These stones were 240,000 years old, but if they had formed on top of the sediments containing *naledi* bones, that placed *naledi* in the cave well before 240,000 years ago, earlier than what we had published. But if *naledi* had broken these flowstones, that placed them here *after* the flowstones formed. In that case, some *naledi* would be younger than we thought. That was an intriguing possibility. Could the species have existed even more recently than we had calculated? These hypotheses would have to be tested. I could feel the weight building of more science that needed to be done.

I opened my orange backpack and began unloading a few of the instruments I wanted to use in Dinaledi. First, I turned on my humidity monitor and took the temperature of the cave: 25°C (77°F) with 10 percent humidity. Then I pulled out my UV flashlight and asked all my companions to extinguish their lights—even the screens on the production crews' cameras. Looking at a cave in black light can reveal hidden things that are often washed out by white light—certain minerals, as well as some fossil bones, fluoresce under a black light.

Once it was pitch-dark, I turned on my UV flashlight and scanned the cave. As the ultraviolet light flooded the walls, bright dots of green and yellow appeared. The stalactites glowed a purplish white, their fluorescence coming from the impurities of their calcite and aragonite crystals.

I started down the steep slope toward the twin passages leading to the Dinaledi Chamber. The way ahead looked tight, smaller than I had imagined. Each of these two passages stretches about six meters long, as I knew from the maps. They run parallel to each other, on either side of a rock pillar that separates their entrances. These tunnels had eluded my understanding ever since the team had first entered the chamber.

As I approached the passages, I thought for the first time that the collective feature was reminiscent of a doorway. As a paleoanthropologist, I have seen thousands of doors in ancient structures. This wasn't a structure, of course—it was a cave. But nevertheless, the two openings side by side evoked a certain kind of transience, a formal movement from one location to another.

That's when I saw the markings.

| 16 |

THE MARKINGS

Over the years I spent staring at the passages connecting the Hill Antechamber and the Dinaledi Chamber in photographs and on the computer screen, it had never occurred to me that they looked like a doorway. But when I was moving through the space myself, it felt undeniable to me that they evoked that kind of feature. Many of the ancient doorways I've seen during the course of my career have borne ancient symbols or signatures that, long ago, shared information about what kind of space lay beyond. Much like our labels on doors for exits, sanctuaries, and even bathrooms, they help us know the function of the space we're entering. So as the doorlike nature of the passage entry struck me, I couldn't deny it. The markings were there on the wall.

I approached the passages. My headlamp's sharp white light illuminated scratches and shapes, squares and triangles, lines that crossed over each other to form shapes resembling the letter *A*. There were crosses—some upside down, others right side up, at least to a modern eye. In the lower half of the rock panel was a square box. The hand, or hands, that had made it clearly attempted multiple times to form the image. A series of

As if decorating an entryway, these markings appear at an opening between the Hill Antechamber and the Dinaledi burial chambers.

ladderlike carvings ran down the left-hand side, and near the top was a shape like a fish with an X in it, the only nongeometric shape I noticed.

THE MARKINGS

Closer inspection shows many figures: lines, crosses, ladderlike strokes, and (at top) a nongeometric shape that looks something like a fish with an X through it.

I couldn't believe what I was seeing. It looked to me as if the markings had clearly been made by hand. They were different from the natural weathering lines that form in hard dolomitic limestone. I was stunned. There should not be engravings in this chamber. Only humans make engravings. And to my knowledge, besides our explorers and a few previous cavers, no humans had ever entered here, certainly none with the time to make engravings like this in the very hard dolomite.

I looked at Dirk. "Did you make these?" I asked. I knew the question was foolish. Certainly, none of our explorers would have defaced a wall like this.

Dirk shook his head. He moved next to me. "Could they be natural lines?" he asked. Dolomite is sometimes called "elephant skin" for the irregular lines that form on it during erosion. But I shook my head in an emphatic no. I have spent the greater part of my adult life clambering in and out of these cave systems. Dolomite develops natural lines when it erodes, but I knew what those lines looked like, and these didn't match. I moved closer to the engravings, pulled out my black light, and turned off my headlamp. Then I shone the black light on the markings.

I expected the black light to create more contrast between the markings and the cave wall, but in my exhausted state, the images seemed to lift from the rock and float before me, shimmering in a brilliant neon blue light as if burned into the air. The weight of what I was seeing landed on my shoulders, and I was nearly overcome with emotion. I had waited most of my life to enter this space, to interact with the species I had been observing from afar for years, and now, thanks to a potent combination of exhaustion, adrenaline, and emotion, I felt untethered from the physical world and transported to their time and space. I had never thought of myself as the sort of person to have visions like this. I was, am, and will always be a skeptic. Yet I can't deny how real it felt, watching those shapes float in front of me.

I flicked off the black light and watched the glowing shapes fade from my optical nerves. It seemed as if I could still see them, floating in space in front of me—an illusion likely caused by my heightened senses and the aftereffects of the black light, something like nighttime headlights dancing in the rearview mirror. The image burned itself into my brain and unnerved me. I shook my head as if to delete the floating illusions, then I turned my headlamp back on. In that bright white light, the engravings reverted into a flat relief. I felt an odd sense of embarrassment as I realized I wasn't alone. I looked around at the others. It had been a deeply personal experience; I didn't want Dirk or anyone else to think something was wrong with me. I took a deep breath to regather my thoughts, then turned back toward the rock to take a closer look at the engravings. The area around them seemed to have been rubbed smooth. It was difficult to fathom how that texture could be achieved through natural processes.

I ran my fingers a millimeter or so from the rock. These shouldn't be here, I thought. *Homo naledi* couldn't have made rock engravings. But then I remembered the conversation with Penny Spikins at Princeton. "They are going to be wondering why there is no rock art on the cave walls," she had said. Maybe we should have expected this. Someone had etched dozens of images into this wall, right above the burials. It had to be *Homo naledi*.

At this point in my story, people often ask me, "How could this be the first time anyone on your team noticed the markings after years of working in Dinaledi?" It's a fair question, and the answer, I think, lies in something I call "backyard syndrome."

Once people become used to a place, they stop observing it so closely. As long as nothing major changes about where they are, people tend to lose small details, such as the exact position of a piece of furniture, or how things are arranged on a shelf. It's a trick of perception, a way for our brains to protect themselves from information overload. It can even be useful to deaden ourselves to unchanging details, especially if it heightens our awareness of new things, potential dangers, or unwelcome changes to the environment. It's the reason you can fall asleep to the sound of a ceiling fan in your bedroom but still wake up to the sound of a car alarm in the street.

Underground, it's easy to develop backyard syndrome. Successive chambers often look similar to one another, and everything is covered with the same coating of dirt. Even on just your second visit to an excavation site, it might take focused concentration to observe new details. In the case of Dinaledi, it's possible that everyone who entered after Steve and Rick in 2013 assumed that everything important had been noticed already—maybe a few people even saw the markings and dismissed them as features that had already been observed and recorded. I don't know. But when I arrived in the Hill Antechamber, I tried to put myself in a state of mind to observe closely and deliberately with fresh eyes.

These carvings were a big deal. The more I looked at them, the more I realized: They were a *huge* deal. I was likely looking at the oldest carvings or etchings of any real complexity yet discovered. The only thing older I could think of were some simple etchings scratched on to a freshwater mussel shell found in Java, Indonesia—simple marks thought to be made by *Homo erectus* perhaps 500,000 years ago. The markings in front of me, however, were anything but simple. They were different geometric shapes, not all alike, and they were abundant. I stood there in the dark surroundings, letting my eyes wander over the panel, taking each image in. All were fairly big: the square almost as large as a playing card, the ladders as long as rulers, and the fishlike shape big enough to be a sticker on a car windshield.

The triangles and X's reminded me of markings found on little pieces of ocher in Blombos Cave, a site almost 1,500 kilometers south, on the coast of the Cape of Good Hope. Those hatch marks and figures, much smaller than these, were often described as the earliest art created by *Homo sapiens,* and they had been dated to about 78,000 years old. But that was the oldest *human* art, I thought, marveling at the figures now in front of my eyes. These remarkable engravings were almost certainly not made by a human. They were right above the burial of a child of *Homo naledi,* deep inside a cave in chambers that *naledi* had used as the last resting place for the dead. These were their markings. They must have meaning. They must represent some sort of deliberate attempt to communicate with other *naledi* who would come afterward—but what were they saying?

Would we ever know the meaning of markings made by a species with a brain a third the size of ours? By another mind?

    I looked at my Mickey Mouse watch. Almost an hour had passed since I had entered the chamber. How was that possible? It felt like mere minutes. I needed to continue exploring.

| 17 |

MORE MARKINGS

I dictated information about the markings into my voice recorder, my voice tinged with a tone of awe, and then forced myself away from this discovery of a lifetime. I entered the passages between the Hill Antechamber and the Dinaledi Chamber and began making my way to Dinaledi. I bent to examine the floor as I went and saw small flecks of bone. There might be burials in the passageway itself, I thought. I began scanning the walls, looking for more engravings. The passage grew narrower than my shoulders, so I turned to fit my upper body. Somewhere close to here, our geologists had mentioned a hole in the side of the passage, a window where you could look down to see a hollow space underneath the whole chamber.

I came to the window and peeked through, shining my headlamp down into a space I could only describe as an abyss. My headlamp couldn't even illuminate the bottom. To the right, I could see that the dirt of the floor I stood on formed a layer that went down for over a meter before it encountered a three-centimeter layer of flowstone that made up the base of the feature, like a thin foundation holding up the floor. It made me queasy to think that thin rock was the only thing holding up the very passage I was standing on. If the rock gave way, we would all plunge into that same darkness my light couldn't penetrate.

I could see why we had never explored this—it looked incredibly dangerous—but the darkness of a cave always contains the allure of a possible secret. There might be a large cavern or chamber acting like a basement underneath Dinaledi. I surmised this expanse had led some of our geologists to think there could be a drain underneath the Dinaledi Chamber itself, sucking the bones downward into formations like the Puzzle Box. This void could be where they thought things vanished. As I looked down into this space, I thought about the relentless downward movement of sediment through the system. Nothing in a cave lasts forever.

The passage narrowed as I continued onward, forcing me into a crawl for the final stretch. If you imagined there was a crawl space behind your bedroom wall, and you squeezed into it and traversed almost the entire side of your home, occasionally crouching down on your hands and knees in order to fit, you'd have an approximation of how it felt to journey to Dinaledi. It was not comfortable. I kept my eyes forward, looking for the telltale sign of blackness swallowing my headlamp to know when the tunnel ended and the expanse of the Dinaledi Chamber began. Finally, I saw my light enter blackness. I paused for a long moment and took a final breath before crawling the last few feet out of the passage. This was the Dinaledi Chamber, the space where this whole adventure had started.

Dinaledi had taken on mythological proportions in my mind, but the chamber itself was in fact bigger than I thought it would be. I shone my light up to the ceiling, so high the edge of my headlamp beam barely brushed the glistening stalactites. Down the length of the chamber, I could see the burial feature in the distance: a raised oval filled with bones, the surrounding dirt excavated away by our team, effectively inverting the grave from a depression into a mound. A wall with large window-size holes in it ran along the center of the chamber, carving the entire space into something resembling a horseshoe. The passage from which I had entered the chamber stood at the base of the horseshoe's right leg.

I hurried to the burials. Pictures and maps could never do justice to the clarity of seeing these graves in real life. I could see where the dirt had actually been moved by *naledi,* how the edges were irregular in shape, and the clumps of dirt that were displaced and then put back into the space. *Naledi* had clearly dug an oval that was larger than the body, part of it filled with disturbed soil. I examined the bones I could identify and imagined how a body would collapse into itself as it began to decay in the grave. The head would fall into the shoulders as the chest collapsed. The humerus at the top of the arm would remain in place, and one end of this bone would stick up above the skull, which would fragment under the weight of time and dirt, and perhaps the trodding of *naledi* feet. The thorax, with its fragile ribs, would crumple like a concertina, and the knees would end up mid-chest. It was a miracle any of this had survived at all, being more than 200,000 years old.

It was becoming more and more difficult to deny that these features were anything but burials, but still, I wanted to do my due diligence and honor our long-held theory of bone flow from Hill into this chamber. I took out my laser measure to take a level reading from both directions: toward the Hill Antechamber and toward the back wall of Dinaledi. What I found stunned me: The laser level read 11 degrees of slope *downward toward the entrance.* The burials were on the high ground. That contradicted most of our maps, which to this point had showed the floor to be practically flat, or even sloping slightly away from the passage I had just exited. But the chamber didn't slope that way; it sloped *toward* the passage.

The evidence was stacking up against the flowing-bones theory. The Chute wasn't a vertical drop, and the bones could not have tumbled into Hill, then into Dinaledi. On top of that, the passages from Hill to Dinaledi created an almost definite choke point that would never conduct the flow of bones or sediment. These bodies certainly weren't swept here from outside the subsystem.

Three hours went by, but I would have told you it was 30 minutes. Making my way back into the center of the Dinaledi Chamber, I looked about. Soon it would be time for me to climb out. I sat down in the middle of the chamber, next to the grave, and began to take the space in one last time, absorbing the moment and preserving every last detail in my mind. I tilted my head up toward the ceiling and opened my eyes just as Dirk, Mathabela, and Warren arrived to collect me. But then I noticed something else.

I pointed up at the ceiling. "Hey, do you guys see that darkened area? And those black dots?"

I pulled out my most powerful flashlight and shone it up at the ceiling. Pristine white stalactites, clearly young formations, had grown over older stalactites that were stained a dark gray. Dark black dots spattered the formations, but this far back in a cave, limestone formations should be a pristine white, not gray and blackened. Something must have stained them, and the most likely culprit I could think of in the moment was fire.

"That looks like soot and smoke damage to me," I said. I had never seen blackening and graying like that by any means other than fire residue. Manganese, a dark metallic mineral often found in these caves, blackened things by mineralizing on surfaces, but in my experience, it didn't look like this. This was more of a general discoloration, an overall graying that, quite simply, looked like smoke staining.

One of the biggest questions that had eluded us to that point was also one of the most basic: How did *naledi* navigate these pitch-black spaces? But if what I was seeing was what I thought it was, the answer had been right above our heads the whole time. *Naledi* had brought fire into these spaces, and that fire's smoke and ash had stained the limestone.

MORE MARKINGS

After another half hour of dictation and directing the others to record the staining, it was finally time for me to climb out of the cave. I moved toward the same passage I had used to enter Dinaledi, the one that would lead me back to the Hill Antechamber, for the return journey. But in a spur-of-the-moment decision, I crouched to approximate the height I thought *Homo naledi* would reach—little more than a meter above the ground. I imagined myself a *naledi* then, slipping easily through the space that, an hour beforehand, I had found uncomfortably tight. As I shuffled along the passage at *naledi* height, I shone my light on the walls. Could they have carried torches at this level? Used small portable flames to light their way? Then I froze. I blinked rapidly. Then I squeezed my eyes shut tight for several moments before opening them again. But I wasn't hallucinating, and I wasn't imagining things.

There, carved on the rock that formed part of the right-hand wall of the passage—the reverse side of the rock with the first set of markings—was an enormous carved crosshatch, reminiscent of a modern-day hashtag, with two more crosses coming off it and an equal sign beside it. I gasped, struggling to gather myself enough to speak. "It's another one!" I cried.

I could clearly see the multiple striations, the repeated groove marks, that formed each line of the figure. This had not been a quick job by whoever had done it. On the Mohs hardness scale, which indicates rocks' hardness on a scale of 1 (soft, like talc) to 10 (hard, like diamond), dolomite is a 4, somewhere between fluorite and quartz, and areas of chert within the dolomite can bring that close to 7. Gouging two-to-three-millimeter scratches into such a rock would take serious effort. It would also take a tool at least as hard, and likely harder, than the rock itself.

I moved closer to the rock face and put on my reading glasses. I wanted to make sure the engraving wasn't natural. The crosshatch seemed a different scale from the markings I had seen earlier; it was larger. As I

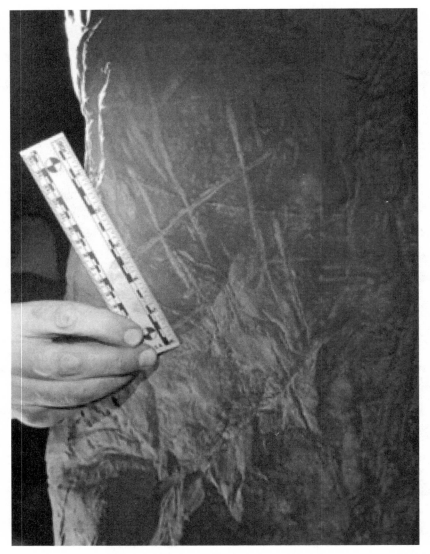

On the back of the same stone column as the first set of markings, another set was found that included clearly deliberate etchings into the cave walls.

examined the lines in the light of my headlamp, I noticed that some of the scratches at the base of the figure crossed over a fossil stromatolite, cutting into the wavy lines of its surface.

Stromatolites are geological deposits that originated as living structures. They formed two and a half billion years ago from microbes that lived in the ancient sea, in an age before complex life evolved. Layer by layer, these simple single-celled organisms built up mineral content and became glued together into microbial mats. This wall had a large cross section of stromatolite that rippled across it like the ridges on a potato chip—and several lines in the crosshatch cut directly through it. One leg of the crosshatch began as a straight line on the smooth dolomite, but when the line reached the stromatolite, the striations began to slip out into several stray lines. I could imagine that the artist had reached the buildup and began struggling to cut straight lines into its irregular surface. To me, it was the clearest indicator that these lines were engraved and not natural.

I noticed other details. Dents or dimples appeared in some areas, as if someone had taken a hammerstone to the surface of the rock. It seemed as if the surface had been sanded, or covered with grit or soil that was the same color as the dirt from the floor. This action, it seemed to me, had not only smoothed and polished the area around the engraving but also filled in some of the etches on the surface. It was as if this part of the pillar, an area roughly 50 by 50 centimeters, was a canvas. It suddenly became clear to me that there were older etchings in the stone that had been either covered by dirt or eroded with age. I could see faint lines underneath, going in directions different from the brighter, more recently carved lines, and it even looked like some of the incisions had been filled with dirt. The crosshatch image that had caught my eye, it seemed, was only the most recent of many carvings no longer as visible. This rock canvas had been marked over a long time, not just in a single event.

As I looked closer at the crosshatch—similar to a tic-tac-toe board or a pound sign—other shapes became clear. There were at least two crosses, maybe three, interlaced with the larger lines. Those resembled the lines in the first set of markings I had seen. There was also what looked like an equal sign formed by two parallel lines to the right of the figure. Cross-hatchings, crosses, and equal signs are all very human images,

CAVE OF BONES

A diagram maps out many non-natural lines on the cave wall, their details suggesting that they were not all made at the same time.

This set of lines, reminiscent of today's hashtag, appears to have been the most recently carved into the cave wall.

created to convey meaning in specific situations. It's tempting to view markings like this through the modern lens of meaning, but if *naledi* had indeed etched these shapes, the artist would surely have different meaning behind it. Whatever these symbols signified or wherever they came from, be it from a single source or a combination of different hands,

I couldn't help thinking of the maker, or makers, as an artist working on a rock canvas.

I spent the next several minutes with Mathabela recording a video, with me pointing out and describing the engravings, trying to capture my impressions in the moment. Then I took a couple of snaps with my phone camera. This broke my pact not to take pictures while in the caves, but I needed to get this for my colleagues waiting outside. Finally, I crawled back into the Hill Antechamber and looked at the pillar again from the other side to compare the new markings to the first set. My eyes passed over the triangles, the square, the ladderlike structures, the crosses, and X's and went up to the fishlike figure higher on the panel. It really was remarkable. After seeing the engraving in the passage, this first one was now even more obviously intentional. One area of the panel, the spot with the fishlike shape, even looked like it had been plastered over with a reddish or yellowish substance. In other places, it seemed as if a white material, perhaps flowstone, had been etched into the lines. And another area looked almost greasy, as if organic material or some other shiny material had been spread over the surface, or as if it had been smoothed by the touching of hundreds of hands, like the shine that develops on the nose of a statue rubbed by believers as they pass by.

I shook my head. This journey had been one of the most momentous experiences of my life—literally transformational. I couldn't unsee the figures floating before me in my mind. It was like the immediate moments following a fireworks show or staring at the sun, when you shut your eyes and the image remains. Surely, these events would affect me for months to come. I would undoubtedly process my experience in Dinaledi over and over again, perhaps for the rest of my life.

I looked around the chamber. It felt as if *naledi* had just left. They had altered this space, and then, when the last one had departed, everything had remained as they had left it. It felt as if we had opened the door to an attic that had been abandoned for decades, old newspapers still turned to a specific page and moth-eaten clothes still hanging, just a covering of dust betraying that time has passed.

But it was time to ascend.

I sat next to the Chute exit and looked up at the ladder and the tiny slot above it that we would all have to pass through on our way out of the cave. A sense of dread bloomed in my stomach. I had no illusions about how hard this climb was going to be. Staring at that tiny slot in the rock three meters or so above my head, I remembered the way it had compressed my chest as I let my body weight and the strength of my arms shove me through on the way down. Now I was going up, and my muscles would need to strain to pull me in the opposite direction. I tried to focus on positive thoughts.

"Let's do this thing," I said.

| 18 |

STRUGGLING OUT

Thirty minutes later, I was in the Chute, wondering whether I would live or die. Through many expeditions and dives, I have faced life-and-death situations. I've been charged by lions and elephants. I've slipped off high cliff walls. I've been stuck in tight spaces for significant periods of time underground.

But this squeeze was the worst situation I had ever been in in my life. I was not only trying to get out alive to see my family and friends again, but I was also carrying observations and information about discoveries that would alter the course of our research for years to come. The symbols, the evidence for fire, the tiny details of the geology that suggested *naledi* had altered the space. I had narrated my findings on the recorder, but I knew there were subtleties and details that wouldn't be conveyed by the audio, or even Mathabela's video. I had to get out to share these revelations with my colleagues, discuss their meaning and importance, and pore over how they could be integrated into our work.

Going down the Chute, as tough as it had been, was easy compared to ascending. I had been fresh and ready when I started out that morning, full of adrenaline. Going up was a different matter entirely. I could tell from the moment I began climbing. The first gap I reached is where I had been forced to compress my chest to drop into the Hill Antechamber. Now I faced a different problem. I had to approach with one arm extended and the other clinging to the wall. This position placed me with just my head, an arm, and a shoulder through the gap before my chest stuck tight. I had expected to use my feet to push myself upward and force myself through with my leg muscles, but the tiny knobs of rock that acted as footholds weren't positioned in a way that I could get myself leverage. I had only one arm to pull myself. It took me almost 10 painful minutes to finally wedge both arms through the gap and slip my sternum above the ledge of rock. I knew that the entire journey would be like this: with my head in one type of squeeze, often with only one of my arms, and with my chest or pelvis contorted in another direction. My legs, meanwhile, would likely be useless, dangling in space or unable to find purchase on the slick walls. This was going to be even harder than I had imagined.

During that first stretch of the climb, I was forced to ask Dirk to knock off another sharp edge of a rock. It was pointed directly at my chest, and I was afraid it would lacerate me. Dirk climbed up to it and knocked off a few centimeters.

After 20 minutes, I was just halfway through the 12-meter journey. I had climbed hardly more than three body lengths. Every centimeter was a titanic struggle, with me relying solely on the strength of my arms. I was out of gas, and even worse, I was stuck. Not just compressed—I was hanging, trapped in a situation where my body simply wouldn't fit through the next tight space.

As my exhausted mind sifted through my options, I thought back to a conversation I'd had at one of the National Geographic Society Explorers Festivals, an annual get-together for adventurers of all kinds: fossil hunters like myself, but also extreme divers, rock climbers, cavers, astronauts, deep-sea explorers, and high-altitude climbers. I've always thought that climbers are the most extreme explorers on Earth. Those individuals push themselves beyond the physical and mental capabilities of most humans. They willingly go into environments that can kill them in a variety of ways—confronting a lack of oxygen at altitude, cold temperatures, avalanches, and exhaustion—and somehow push themselves beyond their limits to survive. Whenever I have a chance to hear their stories at something like the Explorers Festival, I do my best to rein in my tendency to be a chatterbox, and instead I listen.

That evening, a group of explorers had gathered at a bar, and the climbers regaled us with tales of all the extreme things they had seen or experienced on mountainsides—avalanches, crevasses, falls. Johan Reinhard, a famous scientist and climber who had worked on mummies in the high Andes, mentioned surviving a high-altitude ascent by using the "Everest Shuffle." He explained that there is a time in every mountain climber's career when they reach a point of no return: There's nothing left in them; they're gassed. They find themselves staring at their feet with the simple understanding that if they don't slide one foot forward, just to move a half step, then they are going to die on the mountain. That's the Everest Shuffle: one step, then the next, then the next. It's beyond the physical; it's all mental.

I was impressed by Reinhard's story. I had never been to that edge. I'd been tired before, exhausted even, but not to the point that I literally had to move or die. Frankly, it sounded impossible.

But as I hung there halfway up the Chute—not even through the tightest portion yet—I understood the Everest Shuffle. It had taken every single ounce of strength in my body to move six meters, but compared with what was still to come, those six meters had been nothing. And I was stuck anyway. Just as in the Berger's Box, one of my legs was too long to fit where I needed it to go to push myself any farther.

I stared down at my femur, the vile part of my body that had fought me for several minutes. The only way for me to gain purchase was to bring my leg up 90 degrees, but when I tried, my leg wedged itself in the rock, my knee just short of clearing the lip that blocked my way to freedom. I was trapped.

I thought about pulling myself with my one free arm, but from my contorted position, my hand could not find anything I could use for leverage. I also couldn't bring either arm down to push off against the sides of the Chute and lift myself through the gap—I wasn't sure I had the energy to do that anyway. My best course of action was to bring my knee and thigh through the narrow gap I had already cleared up to my chest, then lift my knee over a hard ledge of rock. Once my knee was over this ledge, I could use my thigh to push myself upward—toward the worst squeeze of the Chute, granted, but at least I would be free.

I hung there, gasping. The steam of my breath billowed in the humid air as I looked at the end of my thigh. My whole leg was about two and a half centimeters too long to fit. And no amount of pulling, pushing, tugging, or angling of my hips could make it any shorter. I had made seven or eight attempts to get through, and nothing I could do, no angle or movement, would let the length of my thigh pass. I stared at this spot, memorizing the rock, then my leg. I really didn't have anything left in me; I had reached my Everest Shuffle.

As I looked about, up and around the pinnacle, I could see faint light shining through the series of horizontal cracks from above, probably Maropeng's or Warren's headlamp. I had a vision of *Homo naledi* moving through these passages, coming from all directions down toward the Hill Antechamber below me. They wouldn't have blinked coming through here. They were smaller in stature, slighter in build, and more powerful than me, and had a smaller head. I could visualize them slipping easily through this space, and boy, did I envy their build.

I had run into a terrible situation and had to make a deeply personal decision. The climb to this point had taken everything out of me, and in that moment I had a terrible sense of foreboding that if I relinquished even a centimeter, that I would never get out of Dinaledi. It's hard to engage with that kind of despair. I truly felt there was no way out.

I stared at my knee, and I decided to do something desperate. My only way out of this—the only thing my exhausted, perhaps confused, mind could envision—was to use the ledge to dislocate my patella, slide my knee through the gap, then force my patella back into place. If the maneuver disabled me, well, I would deal with that on the other side of the gap. At least I would be closer to the exit.

I stilled myself by taking deep breaths. I had read about people who cut parts of their bodies off to escape similar circumstances, and I had always tried to imagine where somebody had to be in their head to do that. This was about as close as I was going to get. I couldn't go back. As I stared at my knee wedged below that rock lip, I envisioned its underlying anatomy the way I sometimes envision what lies beneath sediment or behind rock. My patella floated like a turtle shell between the end of my femur and the top of my tibia. Its tendons connect it to those other bones and help it extend the lower leg. But it's a mobile bone and can be dislocated relatively easily. Patellar dislocations happen to athletes all the time, and the bone is actually fairly easy to get back in place. I was rationalizing the idea to myself like a lunatic. In retrospect, I know I wasn't being logical, but situations like this force you into odd decisions.

I found a foothold with my right foot and braced myself against the wall. Then, shutting my eyes and loading up on my right hip, I lunged upward with everything I had and jammed my knee into the rock. Pain coursed down my leg. I groaned. I opened my eyes and saw my kneecap remained stubbornly, safely secure in its proper place. It hadn't worked.

I collapsed back into my uncomfortable position. My leg was literally two and a half centimeters too long, and I couldn't change that fact. Even my most desperate measure had failed. I thought about my family and all the information I had—the discoveries from the chambers. How could I get out of this? I pressed myself into that slot and began trying to push upward in desperate little movements. It was just a couple of centimeters, I kept saying to myself. I twitched and twisted. Minutes went by. But I had nothing left. I remember yelling out loud to myself, "Come on,

After making remarkable discoveries deep in the Dinaledi Chamber, Lee Berger struggles back up the Chute—moments he has called the most difficult in his life. Here, he makes the final reach to extract himself from the Chute's exit.

Berger! Do this!" Dozens of meters away in the Dragon's Back Chamber, John and Kene, as they later told me, could hear my groaning echoing throughout the cave. I was lost in the moment.

What happened next is still a little unclear to me: I remember only pressing and straining, trying to find leverage, and then, after many minutes, I looked down to reset my position to see that the end of my knee had passed over the ledge of rock by perhaps half a centimeter. I stared at this miracle, not understanding it, not trying to understand it. I was free. I had purchase. I pushed my head and arms into the narrow gap above.

I honestly don't remember much about the rest of the journey out. I see it as a series of still images. Maropeng and Warren leaning down in the Chute. The release in my body after passing through the narrow gap. Pulling on the rope. I remember being in the tubular space that made up the last three meters of the climb, a camera, and Maropeng's face. And then I was out.

I lay gasping in the tunnel leading to the Dragon's Back descent. I rolled over and stared back down into that network of fissures, faults, and gaps, down to nearly the bottom. I knew I had pushed myself to my limits, perhaps beyond them. I would never go back down that route again into the Dinaledi Chamber. I felt odd, disassociated from my own body, and as I rested, I settled back into myself.

I pulled myself forward, out of the Chute entrance, and staggered to my knees, then to my feet. I was euphoric, and more exhausted than I had ever been at any other moment of my life. As I reached the bridge—thank goodness now for my safety concerns—my mind filled with the experience of the last few hours and the discoveries. I shakily put on my climbing harness, clipped myself into the steel safety cable, and carried on. Below me, I saw the work lights in the Dragon's Back Chamber. I had made my way into—and more important, out of—Dinaledi. Now I could share this incredible new information.

PART IV

MEANING

| 19 |

MARKINGS AND MEANING

I sat under the olive tree that dominates the sinkhole above the Rising Star entrance and called Jackie. She was nearly in a panic. Our internet had gone out in the cave, and no one had been in a position to call her. I had been in Dinaledi for more than four hours since I had called her, then it took me an hour to climb through the Chute and another hour to move through Dragon's Back. She hadn't heard from me in nearly seven hours. Oops.

After assuring her I was all right, I babbled about the challenges, the discoveries, and the work ahead, and I promised to call with more details. I'm sure I sounded incoherent, but she was a patient listener. She had been on this journey with me. She knew how important this was.

After saying goodbye, I leaned against the olive trunk and watched the group mingling at the entrance of the cave. It looked like a normal end of a day's work, but I felt fundamentally changed by the whole ordeal: the symbols, the visions, the climb, the idea that *naledi* had altered the space, and the evidence of fire. I was a different person than I had been that morning. I didn't know what that difference was exactly,

The crosshatch markings in Dinaledi (top) seem remarkably like those found on the floor of Gorham's Cave in Gibraltar (bottom), identified as being made by a Neanderthal less than 100,000 years ago.

MARKINGS AND MEANING

Tracings of the Dinaledi (top) and Gorham's Cave (bottom) engravings show how strikingly similar they are.

but the journey into those spaces where *naledi* was buried had changed me profoundly.

As the production crew packed up, I walked over to the small group of explorers and scientists. "I need a beer," I said. "And I need to tell you all what I have found." We agreed to assemble at a small pub down the road. It had long wooden benches where we could sit as a group and hash over the day's discoveries. I started for my jeep, but John and Agustín barked at me to stop. John said I looked drunk. Agustín said I looked like I'd had a stroke. I threw them the keys and climbed into the passenger seat.

Just a few of the bar's other tables were occupied on this early weeknight. The crowd was mostly couples chatting and a few small groups of farmers. Over the occasional clink of pool balls coming from the tables nearby, I told the team the story of my journey.

There were so many new impressions to me, things that I hadn't understood before. I had to first communicate that the Chute wasn't a chute at all but a labyrinth of passages through the cave system. I now understood the dirt in Hill was coming from a side passage, not from the Chute itself, and I suspected that *naledi* had broken through flowstones and moved stones about. I was pretty much running a monologue.

Finally, I described the engravings to the group. Halfway through my explanation, I remembered I had taken photos of them on my phone, so I pulled up a picture of the crosshatch engraving and raised the phone for everyone at the table to see.

The reaction I got was not what I had been expecting. Agustín leapt to his feet. "I'll be back!" he cried, and headed toward the parking lot. He looked as if he had just received urgent news. John's reaction puzzled me, too. After glancing at the picture, he became occupied with his phone, typing on it underneath the table like a teenager at the dinner table. I felt a little offended. After all, I had just about died to get these pictures.

Agustín strode back to the table at the same moment John looked up. As if they had coordinated this movement, they both held their phones out for me to see. John's showed a similar crosshatch symbol against what appeared to be a darker rock. Agustín's phone displayed the exact same image. "What is that?" I asked. The symbol in their photos had the same angles of cross-hatching as the main crosshatch I had found in Dinaledi—even the offset cross on the right side of the overall image assumed almost the same position.

"This image was made by a Neanderthal less than 100,000 years ago at Gorham's Cave in Gibraltar," John said.

Agustín nodded. "And it's exactly the same as your image."

I held my phone and John's phone side by side to compare the symbols. It was uncanny. Anyone who looked at these images would suspect they'd been done by the same hand. Yet I knew they came from two different species, one with a brain slightly larger than a chimp's and the other with a brain as big as or bigger than that of a contemporary human. They had been created more than 6,000 kilometers apart, on two different continents, as many as 200,000 years apart in time. Both John and Agustín had been in Gorham's Cave—they knew that engraving from firsthand knowledge.

My mouth hung open as I looked between the images, then at the group. This was mind-blowing. The Rising Star engravings were the first evidence of a truly nonhuman species creating symbols—strikingly similar symbols to what we've attributed to Neanderthals, too. Many scientists have speculated about the meaning and importance of such geometric markings. It's tempting to ascribe meaning to markings like this—the possible *naledi* markings could truly be anything, from a counting system to a star map—but the hard truth is that there's simply no way to know for sure what the markings mean without traveling back in time. Still, the markings' mere existence, and reoccurrence throughout prehistoric times, is significant.

I stared at the photos on my phone. What did a *naledi* individual think while viewing these markings deep beneath the ground? My hallucination of the glowing crisscrossing lines returned. These images were unprecedented, done by a species as morphologically different from living *Homo sapiens* as any sci-fi alien would be from Captain Kirk in an episode of *Star Trek*. In the space of just three days, our research had evolved from a straightforward biological study of *naledi* to an investigation into their behavior and practices. We were nearing something more complex than a picture of a simple animal. We were approaching something like the study of a real nonhuman culture.

And more was on the way.

| 20 |

BURNT BONE

I returned to the Dragon's Back Chamber the next day at the request of Kene. Immediately after my escape from the Chute, she had called me aside to explain that while I had been in Dinaledi, she and her team had found something compelling in the new excavation. I wasn't sure another finding could fit into my overwhelmed brain, but even so, I woke up, clambered down into the cave system, and met Kene by the excavation site.

Kene pointed into the rectangular area of the new excavation, now 10 to 15 centimeters deep. Near the center of the dig site, I could clearly make out a circle of gray discolored clay. Inside the circle, I saw, unmistakably … "Burnt bone and charcoal," Kene said. "And over there"—she pointed to another excavated square of the cave floor—"we have what looks like a large ash layer."

I crouched next to the circle of discolored clay. The material was heat-hardened and clearly ashy—definite signs of an ancient hearth. It looked as if someone at some time had built a small fire and burned meat here, perhaps to eat. I shook my head in disbelief. What were the odds? Yesterday, perhaps at the exact same moment I had spied what seemed to be soot on the roof of Dinaledi, Kene and her team had uncovered separate possible evidence of fire here in the Dragon's Back Chamber. To find just one or the other at such a great time depth as these caverns—some 250,000 years old—would have been extraordinary. But both? It was unbelievable. If we could make a compelling case for tying this evidence to *naledi,* it would skirt many of the challenges others have encountered with confirming fire evidence found in open spaces. These features were small, localized, and deep in a cave system where we could reasonably rule out other sources of fire, such as natural fires from the surface or spontaneous combustion of bat guano. Whatever or whoever made these fires had total control over them.

Another implication leapt to mind. The concept that fire use was exclusive to later hominins was one of the last holdovers of the march-of-progress idea. That assumption allowed archaeologists to feel confident in knowing which species made ancient fires. Considering any traces of fire use that dated between two million and 500,000 years ago, they assumed that *Homo erectus* was the fire maker. Later species—Neanderthals and modern humans—adopted *erectus* fire, inventing more complex uses, like heat-treated tools. Evidence of widespread fire use had been dated to only within the last 300,000 years. No one expected that there was another species making and using fires during any of these time intervals. If we could connect fire to *Homo naledi,* it would throw a massive wrench in this generally accepted time line.

In other words, it was a stereotype at this point that only big-brained hominins had controlled fire, which, before this possible *naledi*-fire connection, had forced us to ask a simple question: How could a small-brained species like *Homo naledi* have survived for a million years or more in the face of bigger-brained, fire-wielding competition? What if the answer was that *naledi* had fire, too? It made sense. *Naledi* had small teeth, often a hallmark of a high-quality diet with lots of cooked foods. *Naledi* lived on landscapes where natural fires raged, on a continent where hominins had likely made fire almost a million years before *naledi* appeared. *Naledi* moved about in caves not that far from Rising Star, at the Swartkrans site, where some of the earliest evidence of the control of fire had been discovered.

I took a closer look at the burnt bones Kene had found. "They're animal bones," she clarified.

I raised my eyebrows at her. "Are you sure?"

She nodded. "They're certainly not *naledi.*"

Working out exactly what animal the bones belonged to would require laboratory analysis. But Kene could see from the bones' size and shape that they didn't belong to *Homo naledi*. Not even close. But that was also the exciting part. Identifying these animals could tell us more about the *naledi* environment.

Obviously, identifying these bones would reveal the species that were present with *naledi,* but it could also point to the types and densities of vegetation around *naledi,* too, based on what we know about those species' respective diets and habitats. We know from their curved fingers that *naledi* are likely to have been climbers, but the anatomy alone can't tell us what they were climbing. We know from the pits on their teeth that the *naledi* diet probably contained some gritty foods, but these features can't tell us about the specific foods they ate. The animal bones opened a potential pathway to learning more about *naledi* by way of their surroundings, the world they lived in daily and the world they experienced outside this cave system. What plants did they eat? What trees did they climb? Did *naledi* hunt? And did the bones come from what they ate? These few bones from Kene and her team were like the first lines of a novel written about the life of *Homo naledi.*

I was buzzing. For almost nine years, John and I, working with our large team, had been piecing together the evidence of what had transpired in Rising Star while *Homo naledi* navigated its caverns and tunnels. Over the course of our research, the story had slowly fallen into place. First, our excavations revealed that *naledi* had been moving and altering things in Dinaledi. Then, we saw that the *naledi* bodies were in holes that had been intentionally dug into the floor of the cave. These observations alone shoved us beyond the realms of anatomy and geology and forced us to begin thinking about *naledi* as a species with potentially complex behavior patterns and a possible culture—a real, possible, nonhuman culture.

Kene's evidence suggested that *naledi* used the Dragon's Back Chamber in a different way than they did the burial chambers of the Dinaledi Subsystem. There were bones here we hadn't found anywhere else; there were fire features here we hadn't found anywhere else, either. The *naledi* world was becoming more complicated. Assembling a complete picture of their world was like cracking a safe, and most of the tumblers required substantial amounts of time and effort to click into place.

But this grand theory of *naledi* culture couldn't be dismissed out of hand. It seemed the height of human arrogance to think that no other species could approach our cultural capacities. I was starting to see that archaeologists could not study an extinct species like *naledi* without building an understanding of its social and environmental reality. We needed a more complete picture of *naledi*. That meant we had to view everything in Rising Star with fresh eyes. This species was capable of much more than we had previously thought.

| 21 |

TRACES OF CULTURE

The western part of the Rising Star cave system houses the titular Rising Star Chamber. We had explored this section only superficially in the past, and while some of our explorers had found fragmentary antelope fossils and even a singular hominin tibia deep in the extremes of the system, the relative riches of Dragon's Back, Dinaledi, and the Hill Antechamber had urged us to focus on those sections instead. We had long considered a sequel expedition back into this remote area to possibly excavate, but it stayed on the back burner amid all the exciting skeletons and burial sites. We hadn't found any *Homo naledi* fossils at all in most of the open parts of this area of the system, including the namesake Rising Star Chamber. My eyes had never noticed anything other than the footprints of cavers.

But if my recent experience in Dinaledi had taught me anything, it's that sometimes your first pass at something can be unreliable. Our team members had passed by the engravings dozens of times—maybe hundreds—without noticing them. And in the Dragon's Back Chamber, we had walked back and forth for years over compelling evidence for fire.

It was worth returning to the Rising Star Chamber just in case we had missed something.

Two days after my journey into Dinaledi, John, Dirk, and I started out for the Rising Star Chamber by way of the northern entrance to the cave system. After skirting tumbled boulders near the entrance and ducking beneath an overhang, we flipped on our headlamps, walked down another shaft, and came to another gnarly passage known as the Upside-Down Turnaround.

This is a psychologically difficult passage. Most people enter its narrow vertical entrance headfirst, pulling themselves through a tight squeeze downward until their helmet hits the floor. You find yourself doing a literal headstand as you wriggle and push until your body contorts itself into a 90-degree angle—a position I was getting used to by this point—and you end up staring down into a small tunnel. Once you commit to making the plunge, there's no going back. It would be extremely difficult, if not impossible, for anyone to pull you out without hurting you.

Dirk went in first. His legs pedaled the air as he descended, like a man being sucked down a tube, before he slid out of sight. I looked at John and gestured for him to go next.

John eyed where Dirk's flailing limbs had vanished, then looked back at me. "I think I'll keep watch from up here," he said. "You guys go on ahead."

I nodded. "Fair."

The way in was as horrible as I remembered. It took me six minutes just to force my body into the tunnel, and once I did and made the turnaround, I faced the heartwarming prospect of crawling for 30 meters along a tight passage while I pushed my backpack in front of me and watched Dirk's feet kicking in the light of my headlamp. The floor of the tunnel was nothing but sharp rocks, and every one of them seemed to poke into me as I crawled by. I hated that tunnel.

Out the other end, Dirk and I journeyed another 200 meters to the Rising Star Chamber, including some slides down slippery slopes and climbs up small passages, until we saw the chamber's telltale stalactites. I caught my breath. It really was beautiful. The stalactites were pristine white, uncontaminated from airborne dust, which lent the chamber the untouched sense of a never-lived-in model home. It was clear from the formation of the rocks and the stalactites that this space had never been close to an entrance to the outside.

"This doesn't feel the same as Dinaledi or Dragon's Back," I noted to Dirk. "It doesn't feel like it's ever been used."

Dirk nodded. "I agree," he said. "It feels almost clean for a cave."

"Which way is Lesedi?" I asked. This chamber and Lesedi shared a roundabout point of connection that our team hardly ever used—it was longer and significantly more difficult than the passage off the Skylight Chamber.

Dirk started toward a slanting rock that angled up at about 40 degrees. "We climb up over this rock, then go along a tunnel and reach the squeeze," he explained. He was already a dozen meters away.

I paused before I followed him. Something on one of the stalactites had caught my eye in the flash of my headlamp. I looked up at the ceiling. Behind a pristine curtain of pure white lime, I saw a gray blackened area stretching up into the older rock on the ceiling. It looked similar to the graying I had seen on the roof of the Dinaledi Chamber, the graying I thought was soot.

"Damn," I muttered. "What if that stuff in Dinaledi was just mineral staining after all?" The Rising Star Chamber just didn't feel occupied or used. If Dinaledi's staining matched the staining here ... A touch of doubt entered my mind. Maybe I was wrong.

Instead of climbing the slab, I advanced into the chamber, almost beneath where Dirk was waiting for me. The angled rock created an open space underneath it, like a small mini-chamber. There was a short step down into the depression, but as I hovered my foot out into space, I stopped and gasped. "Come look at this!" I shouted up to Dirk. "You're not going to believe it!"

"What is it?" he called.

Ash, burnt bone, and charcoal bits on the floor, plus evidence of soot on the walls, suggest the use of fire in the Rising Star Chamber.

The floor was covered with burnt bones, ash, and big chunks of charcoal. This was no tiny hearth like the one in the Dragon's Back Chamber. It was a huge scatter of ash, with clearly identifiable chunks of charcoal and scores of charred animal bones. Had these remains been left by *naledi*?

Dirk hurried over, and we examined the site together. I tried to consider every possible explanation besides made fire: I did not see a single tool or flake of stone from outside the cave. Most of the animal bones appeared to be from small mammals, like rabbits, mice, and rats. There were a few antelope limb bones—no skulls or teeth—and I hadn't seen that combination in any scavenger's or predator's lair, but the bones didn't seem new, either. As we skirted the site, Dirk and I found places where calcite had formed over the ash and parts of the burnt bone. Calcite forms only from deposits left behind by water runoff. These pieces must have been here a very long time for dripping water to form rock on top of them.

Curiously, many of the long bones I saw had spiral fractures on the ends, often created when someone uses a hammerstone to break open a bone to reach the marrow. The assemblage, taken as a whole, seemed like a textbook image of what hunter-gatherers actually eat, the things they catch and kill as they wander the grasslands, and the occasional larger animal that is either scavenged or hunted and killed. It was remarkable.

The charcoal and burnt bone made a short trail that led me into a narrow, low passage I had to navigate by crawling. On either side of me, more tunnels with fire features revealed themselves. Charcoal and ash were everywhere. I felt as if I were crawling into a paleoanthropologist's version of a treasure chamber, with findings and riches beyond my wildest imaginings. I looked ahead, and there in front of me was the crown jewel of the space: a stack of rocks, one of the surest signs of hominin activity.

I crawled closer. Each rock in the stack was the size of an American football, and smaller rocks populated the ground around the stack itself. Thin layers of calcite seemed to have welded the bases of some of the rocks together, and the rocks' surfaces appeared to have been heated and blackened by fire. I examined the base and found ash. All together, these components seemed to make up a very intentional and only somewhat primitive structure. The rocks themselves didn't look like mere fallen roof rock. They had slightly rounded shapes. Could this be a *naledi* hearth still intact? How could I be sure?

I shone my lamp beyond the structure, onto the wall behind it, and

A stack of burnt rocks with burnt bones and ash underneath may have been a hearth, untouched since Homo naledi *inhabited the Rising Star cave system.*

there, stark and obvious, was a small white figure carved into the stone. And right below it was a fragment of broken flowstone. My breath caught in my throat. I rolled over. Behind me, I could clearly see where that exact piece of flowstone had broken away from the roof. The gap shone in my lamp. That shard couldn't have tumbled behind the stack of rocks on its own. Something had moved it. Or something had carried it.

Southern Africa is home to many caves and rock shelters with evidence of human groups who made tools, and archaeologists have attributed many of these findings to Late Stone Age and Iron Age periods. Iron Age people in particular, from about 1200 to 500 B.C., made pottery and used a wide range of iron and ground-stone implements. But the Rising Star system held no evidence of this kind beyond the cave entrances. There was no evidence Iron Age humans had moved into these remote spaces, and there was no evidence of even Late Stone Age humans anywhere in

Above the hearthlike rocks, something like a stick figure has been carved in the wall of the Rising Star Chamber. Features such as this add to the list of details suggesting culture shared by those who traversed these underground passages hundreds of thousands of years ago.

the cave system. I have never seen a Stone Age hearth made by humans without significant stone tool litter around it, so what we had found in this cave seemed very unlike human behavior as we knew it. And the only hominin species we could associate with this area because of geological evidence was *Homo naledi*. This hearth was not biological or anatomical evidence. It was edging into cultural evidence.

If *Homo naledi* was burying bodies in Dinaledi and Lesedi, maybe it used other areas near those chambers for other activities. Maybe *naledi* had dedicated rituals for certain spaces, like cooking in the Rising Star Chamber, or lighting fires in the Dragon's Back Chamber, and places where they ate before or after burying their kin. We had been viewing places like the Rising Star Chamber and Dinaledi as unrelated places where we might find disparate pieces of the *naledi* story, but the hearth told us we needed to study the entire system, all the spaces in tandem, to understand if and how they fit together. The answers on the other side of this study wouldn't just paint a more complete picture of *naledi;* they would make the development of *naledi* culture central to the question of how our ancestors—the hominins who preceded *Homo sapiens*—ultimately became human. If we can find evidence for culture in *naledi,* it could bring us closer to answering how we developed culture ourselves.

| 22 |

THE SEARCH FOR MEANING

In 2015, our team first announced to the world the existence of the new species *Homo naledi* in the journal *eLife*, and in 2017, we published our book *Almost Human*, recounting the discovery of *Australopithecus sediba* and the initial fossil finds from South Africa's Rising Star cave system. At the time we published *Almost Human*, we were only beginning to recognize how this primitive-seeming species challenged our notions of time and place in human evolution. Our discoveries proved to be revolutionary—and controversial. According to conventional thinking, any species with this kind of anatomy should have lived more than two million years in the past. Instead, the *Homo naledi* remains proved they existed about 250,000 years ago. This time line surprised scientists all over the world because it meant that *Homo naledi* must have coexisted with early *Homo sapiens*—the earliest modern humans evolving in Africa at the same time. Humans had not been alone; there was another species with them, surviving and, according to what we found in Rising Star, evidently thriving. Maybe our own species was not as exceptional as we had thought.

Since those initial announcements, our excavations of *Homo naledi* have recovered more than 2,000 bone and tooth fragments, the richest assemblage of ancient human relative fossils ever discovered in Africa. This wealth of material allows us to craft a more vivid physical portrait of this ancestor species than we have seen of most hominin species. *Homo naledi* had a small brain, a frame built for climbing, and a pelvis and trunk, as some of the earliest human relatives did. It had long legs and human-shaped feet, hands that included thumbs suited for toolmaking, and small, human-size teeth.

All traces of the species were found in an unprecedented context, with remains of more than 25 individuals secured in the deepest parts of a complicated, nearly inaccessible underground system that challenges even the most expert cavers today. Our work suggests that *naledi* was a master of this environment, moving through these dark, narrow spaces and using them for different purposes. However controversial the idea may be, the evidence suggests that they used some chambers as special places for their dead—a concept many of our colleagues consider impossible. As we have shared here, however, the evidence takes us even further than that, toward the full realization that this species—though lacking a large brain—manifested cultural complexity of a kind we had never imagined possible in our primitive ancestors. In the past half decade, we have discovered and learned incredible things about their potential capabilities—and those discoveries continue to raise many questions.

We have learned a lot about the anatomy, structure, and inner world of *Homo naledi,* including hints about its social life. But the *naledi* remains within Rising Star are isolated from its broader context—the world outside. Our most recent discoveries, from the markings to the burnt bones to what might be a hearth, hint at everything left to learn. These findings have disrupted so many assumptions we and others have had about ancient human ancestors, and that's made these past few years baffling, exciting, and daunting for those of us involved in the research.

Dozens of scientific papers have been published on *Homo naledi*, reflecting hard work by hundreds of collaborators from all over the world. Among all ancient human relatives, the *naledi* anatomy has become one of the best known to the scientific community. We've accomplished this ubiquity by sharing the shapes and forms of our fossils digitally, so other scientists have been able to study them and incorporate them into their own research. As technology improves, the very nature of the science is changing fast, and our willingness to share through open-access channels has been essential to growing our understanding of this extinct species.

We knew it was controversial to suggest *Homo naledi* buried its dead. Now we are suggesting an organized lifestyle for this small-brained, enigmatic hominin. We discovered a stone shaped like a tool near the hand of a buried *naledi* child. Could this be evidence of ritual burial with artifacts, something we thought only humans did? We found evidence of fire in several cave locations, including soot on the walls and ceilings, charcoal, burnt animal bones, and piled stones that suggest a hearth. These findings also point to a differential use of spaces, with specific sites used for burials and other spaces used for cooking animals. Most remarkable of all, certain walls in the Dinaledi Chamber and other spaces bear etchings in the rock—lines that could not have been created naturally, and shapes that inexplicably mirror those found in other caves thousands of kilometers away, caves occupied by species of ancient human relatives with larger brains than those of *Homo naledi.* Stepping back, we are beginning to understand that all these features combine to suggest what anthropologists might define as "culture."

Continued investigation found that a total of *three* walls in the Dinaledi Chamber contain markings. Most are geometric shapes: triangles, hatch marks, crosses, and squares. Other carvings resemble ladders, horizontal lines crossing triangles that make something like the letter *A,* and even a fish with an X slashed through its inside. It's tempting to interpret all these images, but it's essential to remember that the mind that made them was different from ours, both biologically and culturally. We can't project our interpretations onto *naledi*.

We have to eventually confirm the markings' age. That is going to be a challenge several years in the undertaking. It's difficult to date wall markings because removing rock by etching it does not change anything about its chemistry. But there is hope in the science. Recently, geochronologists have been able to estimate the age of some ancient wall carvings by testing layers of calcite that have built on top of them. This kind of approach has dated a painted image of a wild pig in Borneo to around 40,000 years ago—the oldest known image of *figurative* art known to humans. The potential to assign an absolute age to the markings made in the Dinaledi Chamber, and elsewhere in Rising Star, is growing, but the project, too, will likely go on for many years to come. Colleagues, including Agustín Fuentes, have built databases of similar objects from around the world—including the Gorham's Cave marking, made by Neanderthals less than 100,000 years ago. But the markings in the Rising Star cave system might be far older, perhaps 240,000 years of age to be contemporary with *naledi*.

What could explain this common set of images in the first place? Rock art specialists have wondered this for a long time. Some speculate that this tendency to make geometric patterns was fundamental to the development of symbolic thinking, even the beginnings of mathematics or language. These ideas are hard to test. We have only modern humans as examples of a symbol-using nervous system. That makes it hard to see what elements of symbolic thinking come from development, genetic inheritance, or learning.

Still, the markings give us what we need to begin reconstructing *naledi* culture. The Rising Star cave system is like a *naledi* spaceship, access to an alien world in the midst of our own. We have cracked the spaceship's sealed doors open and crawled inside, searching for artifacts and clues about how these peculiar beings used this space, a space that possibly was so important to them that they left their dead in its many passages.

We have faced serious skepticism on this journey. When we were working in the tracks of our past experience—for example, classifying *naledi* within the genus *Homo*—nobody published any critical comments. But as we have moved into the cultural realm—for example, suggesting that *naledi* undertook deliberate burial practices—reactions have been intense. Those who accepted the idea that *naledi* deliberately disposed of their dead tended to leap way past us, concluding that this behavior must mean that *naledi* itself was human. After all, we had never studied a species with this degree of cultural complexity that wasn't us. Archaeologists had believed for decades that mortuary practices, such as burial of the dead, were diagnostic of modern human culture.

But considering *naledi* to be human just because it was complex seems to us an intellectual cop-out. Saying *naledi* was human just avoids the question of how we became human in the first place. And it assumes that any species with complex behavior must by definition be just like us.

Both *naledi* and humans evolved from common ancestors that lived deep in the past. Did both species inherit cultural proclivities from our ancient common ancestors? Or did we evolve those proclivities independently? Or, more compelling still, could the genetic origins of culture have emerged through interaction *between* our two lineages? These interrogations seemed so much more worthwhile and valuable than saying *naledi* was human. From our perspective, that simplification comes from a bias that says humans can only interact with other humans, learn from other humans, or be descended from other humans. It's really a way of saying that in the story of evolution, only humans matter.

Why does this point of view hang on? The idea that humans are the apex of evolution has been a strong current through the history of evolutionary science. Even as anthropologists acknowledge that we are one branch of a complex tree, their quest has been to understand why our branch is so special. The concept of a march of progress keeps snapping back into the picture. We have to recognize this in order to set it aside and consider the evidence in front of us.

We face a deep resistance, in other words, to the idea that *naledi* could be complex in any way that might jeopardize human uniqueness and the underlying assumption that our evolution was a march of progress, with every species of human ancestor in its proper place in the line of evolution culminating in *us*.

It's an intriguing question to think about the comparison, and possible interaction, of our species, *Homo sapiens,* with theirs, *Homo naledi.* The dating of our fossil finds to 200,000 to 300,000 years ago certainly opens up the possibility that the two species coexisted in Africa.

By almost any definition, *Homo naledi* is not human. But if the present archaeological record reflects the complexity of *Homo sapiens* accurately, it means that *naledi* was significantly more complex than *sapiens* at the time. We are likely finding evidence of *naledi* practicing mortuary rituals and creating meaningful symbols as much as 100,000 years before evidence shows humans doing the same things—likely in the same geographical locations.

That is remarkable to contemplate. Imagine a group of small-brained but culturally complex *naledi* meeting larger-brained but less culturally sophisticated humans on the Highveld. Would the interaction be violent? Peaceful? Would the species avoid one another, as chimpanzees and gorillas do today? Or would *naledi* willfully exclude *Homo sapiens* from the regions it occupied?

We may one day find out answers to all of these questions. Genetics or ancient proteins may even reveal that an introgression—a sharing of genes—occurred between *naledi* and humans. If it did, was it the spark that led to the leap in evolution often called the modern human revolution? Did our large-brained ancestors mate with a small-brained but smarter species, and did that union create the magic moment that allowed us to harness the potential of our big brain?

Or, alternatively, was the development of *Homo naledi* and its cultural behavior just a different evolutionary experiment in intelligence? Perhaps *naledi* represents the earliest experiment in a human level of intelligence we have yet discovered, and then our own lineage independently made similar yet separate advances tens of thousands of years later. Another possibility, of course, is that humans saw the practices of *naledi* and mimicked them, developing cultural complexity by observation. Would it not be ironic if the development of what we have so long considered uniquely human culture came about through either interspecies sex or mimicry? Or what if human-level intelligence is just a natural process of evolutionary complexity in our lineage as well as others that has happened many times, and we have just missed that until now?

The possibilities are many, and we must remain open to all of them. At this stage, we simply don't have the answers to these interesting questions. But as the science and technology improve, even down to the molecular level, and as further exploration and excavations proceed, we will almost certainly gain more evidence about the fascinating culture of *Homo naledi* and others in our human family tree.

We've long thought that communication, cooperation, and social control of emotions like fear and trust are behavioral traits humans almost exclusively enjoy. But just as the evidence of such behavior in elephants, dolphins, octopuses, and even honeybees posits, so does our experience with *naledi* argue that these traits by themselves cannot define humans. We don't think this diminishes the importance of these qualities; in fact, we believe it reinforces how central they are to humanity. These behaviors lie in who we are today, and they're part of the story of where we came from. If anything, this is more reason to celebrate our world's extraordinary resilience, complexity, and diversity. Paleoanthropology and related fields like archaeology and history aren't designed to tell us how we became separate and unique from nature. They exist to reveal the wondrous mechanisms of the natural world, to reveal the secrets of our origins, and to bind us to our ancestry. They do not restrict our progress, but rather help us understand what makes us who we are and help us preserve those essential, beautiful parts of being *Homo sapiens*.

By defining *naledi,* we can define humans.

EPILOGUE

At home in South Africa, I often sit at my favorite rooftop bar in Bedfordview and look out over the green expanse of Johannesburg. The view from 12 stories up is magnificent. In the distance, the eight cooling towers of a local power plant trail white steam. Behind them, jumbo jets take their final approach to OR Tambo International Airport. This rooftop has become my favorite place to work, an indoor-outdoor office.

I open my laptop for a call with Enrico Cappellini, Alberto Taurozzi, and Palesa Madupe, collaborators at the University of Copenhagen who are experts in paleoproteomics, a new area of research analyzing the proteins from ancient teeth and bone. Protein is more stable than DNA over time, and so this new technology offers a fresh way to study fossils. Enrico has recovered protein from a hominin fossil found in Spain, dating it back more than 780,000 years. It's an astounding number, but this new approach allows us to potentially date fossils going back *millions* of years. Protein data might also provide information about the sex of ancient fossils such as teeth—mammalian X or Y chromosomes can show up in enamel proteins.

I hope to use paleoproteomics on *naledi* and *sediba*. Any information at all about the temporal relationships of these species would be useful for our work. And our *naledi* findings included so many individuals. A way to determine the sex of all of them would give us valuable data about the composition of *naledi* social groups and how they used the cave system. I gave these scientists two *sediba* teeth and four *naledi* teeth representing different individuals, and I am eager to hear what their analysis found.

"We are getting very interesting results from the proteins," Enrico says. "The dentin has good collagen preservation in our initial samples."

This is terrific news. If the collagen is in good condition, it could mean the fossils are well enough preserved to move to areas of analysis beyond sampling for ancient proteins. "Does that mean there might be good-enough preservation to also get DNA?" I ask.

Enrico nods. "I think it may be possible, yes."

I smile. Here we go, I think to myself. I can already see the next episode unfolding.

ACKNOWLEDGMENTS

This book evolved over the course of five years, across our various expeditions and laboratory analyses into the world of *Homo naledi* and the Rising Star cave system. Our greatest debt, therefore, is to the many explorers and technicians who risked life and limb in order to make discoveries. You pushed the limits of your physical abilities and challenged yourselves to produce potentially world-changing work. Thank you.

We are also indebted to the more than 100 scientists and students who worked with us to produce an extraordinary amount of scientific research about the anatomy, physiology, and context of *Homo naledi*. Your insights, counterpoints, and questions pushed us to stay curious and open-minded. It must be said, as well, that the opinions and personal reflections on the discoveries and results in this book are wholly our own.

The members of the Hill exploration team and the colleagues who ventured into the most difficult and dangerous spaces to recover data deserve special mention: Rick Hunter, Steven Tucker, Marina Elliott, Becca Peixotto, Lindsay Eaves Hunter, Hannah Morris, Elen Feuerriegel, Alia Gurtov, Dirk van Rooyen, Ashley Kruger, Zoë Rosen, Eric Roberts, Maropeng Ramalepa, Tebogo Makhubela, Mathabela Tsikoane, Corey Jaskolski, Kenny Broad, Keneiloe Molopyane, Kerryn Warren, Angharad Brewer-Gillham, Zubiar Jinnah, and Samuel Nkwe. We also thank the Speleological Exploration Club of South Africa for assisting us on larger expeditions, mapping, safety, and security on-site. We are forever grateful for your collaboration and enthusiasm for exploration.

Our support staff has always helped keep the offices going and the fieldwork running seamlessly. Special mention must go to the efforts of Bonita de Klerk, Justin Mukanku, and Sonia Sequeira, as well as to the field staff at Rising Star, including Molly Johnson, Samuel Makinita, Irene Maphosa, Piet Matshinise, Tumelo Molefyane, Zandile Ndaba, and Danny Mithi. For security and maintenance on-site, we thank Gift

Otinela, Samedo Petrus, and Lerato Phiri. The casting team stationed at the University of the Witwatersrand, composed of Bongani Nkosi and Mduduzi Nyalunga, provided invaluable support for creating high-quality scientific replicas for studying the anatomy of *naledi* and communicating the science with colleagues and the public. Chris Collingridge risked life and limb to take photographs of the symbol in the chamber.

There are too many volunteers and students who have helped us on-site and in the laboratory to mention individually, but we recognize that without their added assistance, our work would be much more difficult.

Lisa Thomas of National Geographic Books has supported us since the very conception of this book. You are an enthusiastic supporter, reader, and editor. Our devoted book editors, Susan Hitchcock and Tyler Daswick, worked significant and tireless hours to produce this book in near record time. Thanks to National Geographic's amazing art and design team as well: Sanaa Akkach, Lisa Monias, Mike McNey, Jason Treat, Elisa Gibson, Meredith Wilcox, and Adrian Coakley.

Our donors, sponsors, and supporters make the work described in this book possible. Worth special mention among these are Lyda Hill and the National Geographic Society, as well as the trustees of the Lee R. Berger Foundation: Mark Read, James Hersov, Jane Evans, Jackie Berger, and William Haseltine. Thank you for your financial support, and to the trustees for allowing us access to the site.

The South African National Heritage Agency issues the permits to conduct scientific work at Rising Star and the permits for fossils to travel and be sampled. We are grateful for their continued support of our work.

Our respective universities, the University of the Witwatersrand and the University of Wisconsin–Madison, have given us the freedom to pursue this academic endeavor, and for this we are tremendously appreciative. The former provides curatorial support, and we are grateful to the university curators Bernhard Zipfel and Sifelani Jirah for their constant assistance with laboratory work.

ACKNOWLEDGMENTS

Our families have endured weeks without us at home. Jackie earned a Ph.D. in paleoanthropology just to work closer with Lee, and both Matthew and Megan have been enthusiastic participants in their father's fieldwork at Rising Star. Without their unflagging and enthusiastic support, participation, and love for science, this work just wouldn't happen. John's wife, Gretchen, and children, Sophie, Lucy, Sadie, and Goodwin, have been amazing supporters as well. They are a constant source of insights that have inspired more great science than any of them know.

Thank you, all. Never stop exploring!

Appendix A: Known Humans Who Have Entered the Dinaledi Chamber

(In approximate order of entry)

1. Neil Ringdahl
2. Rick Hunter
3. Steven Tucker
4. John Dickie
5. Selena Dickie
6. Bruce Dickie
7. Matthew Dickie
8. Matthew Berger
9. Megan Berger
10. Marina Elliott
11. Becca Peixotto
12. Lindsay Eaves Hunter
13. Hannah Morris
14. Elen Feuerriegel
15. Alia Gurtov
16. Christo Saayman
17. Pieter Theron
18. Andre Doussy
19. Allen Herweg
20. Michael Herweg
21. Rupert Stander
22. Lindin Mazilis
23. Dirk van Rooyen
24. Ashley Kruger
25. Zoë Rosen
26. Garrreth Bird
27. Eric Roberts
28. Maropeng Ramalepa
29. Elliott Ross
30. Tebogo Makhubela
31. Mathabela Tsikoane
32. Riaan Hugo
33. Corey Jaskolski
34. Kenny Broad
35. Juan Luis Arsuaga
36. Ignacio Martínez Mendizábal
37. Carlos Lorenzo Merino
38. Rolf Quam
39. Keneiloe Molopyane
40. Kerryn Warren
41. Angharad Brewer-Gillham
42. Raymond Messitar-Tooze
43. Zubiar Jinnah
44. Samuel Nkwe
45. Warren Smart
46. Lee Berger
47. Ginika Ramsawak
48. Sarah Johnson
49. Chris Collingridge

Appendix B:
Time Line of *Homo naledi* Discoveries

	2013
September	Discovery of Dinaledi Chamber by Rick Hunter and Steve Tucker
November	First field expedition Discovery of Lesedi Chamber
	2014
February	Lesedi Chamber exploration
May	Workshop to describe and name *Homo naledi*
	2015
February	Lesedi Chamber excavation (intermittent through 2016)
September	Worldwide announcement of *Homo naledi* Research on *Homo naledi* published in *eLife*
	2016
April	Scientific symposium presenting anatomical work on *Homo naledi*
May	Publication of phylogenetic work on *Homo naledi*
July	Establishment of base camp at Rising Star
October	Study of Lesedi Chamber fossils Varied test results show date of Dinaledi Chamber to be younger than expected
	2017
March	*eLife* accepts papers on Lesedi research and the age of Dinaledi
May	Worldwide announcement of Lesedi Chamber and Neo *Almost Human* published
July	Publication of descriptive papers of *Homo naledi* anatomy

APPENDIX B

	2017 *continued*
September	Expedition in Hill Antechamber and Lesedi Names assigned to Hill Antechamber and Dinaledi Subsystem Possible discovery of burial feature in Hill Antechamber Discovery of Letimela skull
	2018
March	Expedition to Hill Antechamber to extract burial feature in plaster jacket
May	Publication of *Homo naledi* brain study
November	Expedition to Dinaledi Discovery of burial features in Dinaledi
	2019
January	Full LIDAR/photogrammetric survey of Dinaledi and Hill
May	Study of Letimela skull and Dinaledi infant begin
October	Unveiling of Neo artist reconstruction
	2020
January	Segmentation of Hill Antechamber burial feature complete
May	Laboratory operations relocated to Rising Star
	2021
October	Worldwide release of Letimela research
	2022
January	Preparation of research describing burials in Hill and Dinaledi
March	Completion of lab work on burials
June	Meeting at Princeton University and finalization of burial papers
July	Dragon's Back Chamber expedition Lee Berger descends to Dinaledi and Hill Discovery of fire evidence in cave system

2022 *continued*	
July	Discovery of engravings in Hill
	Discovery of hearth in Rising Star Chamber
August	Submission of burial research for peer review
December	Carnegie Institution of Science lecture announcing fire discovery

BIBLIOGRAPHY

Bailey, S. E., et al. "The Deciduous Dentition of *Homo naledi:* A Comparative Study." *Journal of Human Evolution* 136 (2019): 102655.

Berger, L. R., and M. Aronson. *The Skull in the Rock: How a Scientist, a Boy, and Google Earth Opened a New Window on Human Origins.* National Geographic Books, 2012.

Berger, L. R., D. J. de Ruiter, S. E. Churchill, et al. "*Australopithecus sediba:* A New Species of Homo-like Australopith From South Africa." *Science* 328, no. 5975 (2010): 195–204.

Berger, L. R., and J. D. Hawks. *Almost Human: The Astonishing Tale of* Homo naledi *and the Discovery That Changed Our Human Story.* National Geographic Books, 2017.

Berger, L. R., J. D. Hawks, D. J. de Ruiter, et al. "*Homo naledi,* a New Species of the Genus *Homo* From the Dinaledi Chamber, South Africa." *eLife* 4 (2015): e09560.

Berger, L. R., J. D. Hawks, P. H. Dirks, et al. "*Homo naledi* and Pleistocene Hominin Evolution in Subequatorial Africa." *eLife* 6 (2017): e24234.

Berger, L. R., and B. Hilton-Barber. *In the Footsteps of Eve: The Mystery of Human Origins.* National Geographic Books, 2000.

Berthaume, M. A., L. K. Delezene, and K. Kupczik. "Dental Topography and the Diet of *Homo naledi.*" *Journal of Human Evolution* 118 (2018): 14–26.

Bolter, D. R., and N. Cameron. "Utilizing Auxology to Understand Ontogeny of Extinct Hominins: A Case Study on *Homo naledi.*" *American Journal of Physical Anthropology* 173, no. 2 (2020): 368–80.

Bolter, D. R., M. C. Elliott, et al. "Immature Remains and the First Partial Skeleton of a Juvenile *Homo naledi,* a Late Middle Pleistocene Hominin From South Africa." *PLOS One* 15, no. 4 (2020): e0230440.

Bower, B. "Pieces of *Homo naledi* Story Continue to Puzzle." *ScienceNews,* April 19, 2016.

Bowland, L. A., et al. "*Homo naledi* Pollical Metacarpal Shaft Morphology Is Distinctive and Intermediate Between That of Australopiths and Other Members of the Genus *Homo*." *Journal of Human Evolution* 158 (2021): 103048.

Brophy, J. K., et al. "Comparative Morphometric Analyses of the Deciduous Molars of *Homo naledi* From the Dinaledi Chamber, South Africa." *American Journal of Physical Anthropology* 174, no. 2 (2021): 299–314.

Brown, P., et al. "A New Small-Bodied Hominin From the Late Pleistocene of Flores, Indonesia." *Nature* 431, no. 7012 (2004): 1055–61.

Christie, G. P., and D. Yach. "Out of Africa: From *Homo naledi* to '*Homo cyborg*.'" *South African Journal of Science* 112, no. 1–2 (2016): 1.

Cofran, Z., C. VanSickle, et al. "The Immature *Homo naledi* Ilium From the Lesedi Chamber, Rising Star Cave, South Africa." *American Journal of Biological Anthropology* 179, no. 1 (2022): 3–17.

Cofran, Z., and C. S. Walker. "Dental Development in *Homo naledi*." *Biology Letters* 13, no. 8 (2017): 20170339.

Davies, T. W., et al. "Distinct Mandibular Premolar Crown Morphology in *Homo naledi* and Its Implications for the Evolution of *Homo* Species in Southern Africa." *Scientific Reports* 10, no. 1 (2020): 1–3.

de Ruiter, D. J., et al. "*Homo naledi* Cranial Remains From the Lesedi Chamber of the Rising Star Cave System, South Africa." *Journal of Human Evolution* 132 (2019): 1–4.

Dirks, P. H., L. R. Berger, et al. "Geological and Taphonomic Context for the New Hominin Species *Homo naledi* From the Dinaledi Chamber, South Africa." *eLife* 4 (2015): e09561.

Dirks, P. H., E. M. Roberts, et al. "The Age of *Homo naledi* and Associated Sediments in the Rising Star Cave, South Africa." *eLife* 6 (2017): e24231.

Durand, F. "*Naledi*: An Example of How Natural Phenomena Can Inspire Metaphysical Assumptions." *HTS: Theological Studies* 73, no. 3 (2017): 1–9.

Dusseldorp, G. L., and M. Lombard. "Constraining the Likely Technological Niches of Late Middle Pleistocene Hominins With *Homo naledi* as Case Study." *Journal of Archaeological Method and Theory* 28, no. 1 (2021): 11–52.

Elliott, M. C., et al. "Description and Analysis of Three *Homo naledi* Incudes From the Dinaledi Chamber, Rising Star Cave (South Africa)." *Journal of Human Evolution* 122 (2018): 146–55.

Feuerriegel, E. M., D. J. Green, et al. "The Upper Limb of *Homo naledi*." *Journal of Human Evolution* 104 (2017): 155–73.

Feuerriegel, E. M., J. L. Voisin, et al. "Upper Limb Fossils of *Homo naledi* From the Lesedi Chamber, Rising Star System, South Africa." *PaleoAnthropology*, 2019: 311–49.

Friedl, L., et al. "Femoral Neck and Shaft Structure in *Homo naledi* From the Dinaledi Chamber (Rising Star System, South Africa)." *Journal of Human Evolution* 133 (2019): 61–77.

Garvin, H. M., et al. "Body Size, Brain Size, and Sexual Dimorphism in *Homo naledi* From the Dinaledi Chamber." *Journal of Human Evolution* 111 (2017): 119–38.

Guatelli-Steinberg, D., et al. "Patterns of Lateral Enamel Growth in *Homo naledi* as Assessed Through Perikymata Distribution and Number." *Journal of Human Evolution* 121 (2018): 40–54.

Harcourt-Smith, W. E., et al. "The Foot of *Homo naledi*." *Nature Communications* 6, no. 1 (2015): 1–8.

Hawks, J. "The Latest on *Homo naledi:* A Recent Addition to the Human Family Tree Doesn't Fit in Clearly Yet." *American Scientist* 104, no. 4 (2016): 198–201.

Hawks, J., and L. R. Berger. "The Impact of a Date for Understanding the Importance of *Homo naledi*." *Transactions of the Royal Society of South Africa* 71, no. 2 (2016): 125–28.

Hawks, J., M. Elliott, et al. "New Fossil Remains of *Homo naledi* From the Lesedi Chamber, South Africa." *eLife* 6 (2017): 6.

Herce, R. "Is *Homo naledi* Going to Challenge Our Presuppositions on Human Uniqueness?" In *Issues in Science and Theology: Are We Special?*, edited by M. Fuller et al., 99–106. Springer, 2017.

Holloway, R. L., et al. "Endocast Morphology of *Homo naledi* From the Dinaledi Chamber, South Africa." *Proceedings of the National Academy of Sciences* 115, no. 22 (2018): 5738–43.

Irish, J. D., S. E. Bailey, et al. "Ancient Teeth, Phenetic Affinities, and African Hominins: Another Look at Where *Homo naledi* Fits In." *Journal of Human Evolution* 122 (2018): 108–23.

Irish, J. D., and M. Grabowski. "Relative Tooth Size, Bayesian Inference, and *Homo naledi*." *American Journal of Physical Anthropology* 176, no. 2 (2021): 262–82.

Kivell, T. L., et al. "The Hand of *Homo naledi*." *Nature Communications* 6, no. 1 (2015): 1–9.

Kruger, A., and S. Badenhorst. "Remains of a Barn Owl *(Tyto alba)* From the Dinaledi Chamber, Rising Star Cave, South Africa." *South African Journal of Science* 114, no. 11–12 (2018): 1–5.

Kupczik, K., L. K. Delezene, and M. M. Skinner. "Mandibular Molar Root and Pulp Cavity Morphology in *Homo naledi* and Other Plio-Pleistocene Hominins." *Journal of Human Evolution* 130 (2019): 83–95.

Laird, M. F., et al. "The Skull of *Homo naledi*." *Journal of Human Evolution* 104 (2017): 100–123.

Langdon, J. H. "Case Study 16: Democratizing *Homo naledi*: A New Model for Fossil Hominin Studies." In *The Science of Human Evolution*, 123–32. Springer, 2016.

Lents, N. H. "Paleoanthropology Wars: The Discovery of *Homo naledi* Has Generated Considerable Controversy in This Scientific Discipline." *Skeptic* 21, no. 2 (2016): 8–12.

Marchi, D., et al. "The Thigh and Leg of *Homo naledi*." *Journal of Human Evolution* 104 (2017): 174–204.

Morwood, M. J., P. Brown, et al. "Further Evidence for Small-Bodied Hominins From the Late Pleistocene of Flores, Indonesia." *Nature* 437, no. 7061 (2005): 1012–17.

Morwood, M. J., R. P. Soejono, et al. "Archaeology and Age of a New

Hominin From Flores in Eastern Indonesia." *Nature* 431, no. 7012 (2004): 1087–91.

Nilsson Stutz, L. "Embodied Rituals and Ritualized Bodies: Tracing Ritual Practices in Late Mesolithic Burials." *Acta Archaeologica Lundensia* 46 (2003).

Odes, E. J., et al. "A Case of Benign Osteogenic Tumour in *Homo naledi:* Evidence for Peripheral Osteoma in the UW 101-1142 Mandible." *International Journal of Paleopathology* 21 (2018): 47–55.

Pettitt, P. "Did *Homo naledi* Dispose of Their Dead in the Rising Star Cave System?" *South African Journal of Science* 118, no. 11–12 (2022).

Randolph-Quinney, P. S. "The Mournful Ape: Conflating Expression and Meaning in the Mortuary Behaviour of *Homo naledi*." *South African Journal of Science* 111, no. 11–12 (2015): 1–5.

Randolph-Quinney, P. S., et al. "Response to Thackeray (2016)—The Possibility of Lichen Growth on Bones of *Homo naledi:* Were They Exposed to Light?" *South African Journal of Science* 112, no. 9–10 (2016): 1–5.

Robbins, J. L., et al. "Providing Context to the *Homo naledi* Fossils: Constraints From Flowstones on the Age of Sediment Deposits in Rising Star Cave, South Africa." *Chemical Geology* 567 (2021): 120108.

Schroeder, L., et al. "Skull Diversity in the *Homo* Lineage and the Relative Position of *Homo naledi*." *Journal of Human Evolution* 104 (2017): 124–35.

Skinner, M. F. "Developmental Stress in South African Hominins: Comparison of Recurrent Enamel Hypoplasias in *Australopithecus africanus* and *Homo naledi*." *South African Journal of Science* 115, no. 5–6 (2019).

Stringer, C. "Human Evolution: The Many Mysteries of *Homo naledi*." *eLife* 4 (2015): e10627.

Thackeray, F. J. "Estimating the Age and Affinities of *Homo naledi*." *South African Journal of Science* 111, no. 11–12 (2015): 1–2.

Tönsing, D. L. "Homo Faber or Homo Credente? What Defines Humans, and What Could *Homo naledi* Contribute to This Debate?" *HTS Theological Studies* 73, no. 3 (2017): 1–4.

Towle, I., J. D. Irish, and I. De Groote. "Behavioral Inferences From the High Levels of Dental Chipping in *Homo naledi*." *American Journal of Physical Anthropology* 164, no. 1 (2017): 184–92.

Traynor, S., M. Banghart, and Z. Throckmorton. "Metatarsophalangeal Proportions of *Homo naledi*." *South African Journal of Science* 115, no. 5–6 (2019): 1–8.

Traynor, S., D. J. Green, and J. Hawks. "The Relative Limb Size of *Homo naledi*." *Journal of Human Evolution* 170 (2022): 103235.

Ungar, P. S., and L. R. Berger. "Brief Communication: Dental Microwear and Diet of *Homo naledi*." *American Journal of Physical Anthropology* 166, no. 1 (2018): 228–35.

VanSickle, C., et al. "*Homo naledi* Pelvic Remains From the Dinaledi Chamber, South Africa." *Journal of Human Evolution* 125 (2018): 122–36.

Walker, C. S., et al. "Morphology of the *Homo naledi* Femora From Lesedi." *American Journal of Physical Anthropology* 170, no. 1 (2019): 5–23.

Williams, S., et al. "The Vertebrae and Ribs of *Homo naledi*." *Journal of Human Evolution* 30, no. 1 (2016): e19.

Williams, S., et al. "The Vertebrae and Ribs of *Homo naledi*." *Journal of Human Evolution* 104 (2017): 136–54.

Wong, K. "Debate Erupts Over Strange New Species: Skeptic Challenges Notion That Small-Brained *Homo naledi* Deliberately Disposed of Its Dead." *Scientific American*, April 8, 2016.

ILLUSTRATIONS CREDITS

Interior and insert images are courtesy of Lee Berger unless otherwise noted below.

Cover, courtesy of John Hawks; 26–7, Private Collection/Photo © Leonard de Selva/Bridgeman Images; 29–38, courtesy of John Hawks; 50–3, courtesy of John Hawks; 55, Stefan Fichtel/National Geographic Image Collection; 58, sculpture by John Gurche, photo by Mark Thiessen, NGP; 61, Jon Foster/National Geographic Image Collection; 73 and 79, courtesy of John Hawks; 98, courtesy of John Hawks.

Photo insert: 1–2 Robert Clark/National Geographic Image Collection; 3, sculpture by John Gurche, photo by Mark Thiessen, NGP; 5 (LO), Elliot Ross/National Geographic Image Collection; 6 (UP), John Gurche/National Geographic Image Collection; 6 (LO)–7, Stefan Fichtel/National Geographic Image Collection; 8 (UP), Daniel Born/GreatStock/Science Photo Library; 8 (LO), Rachelle Keeling.

INDEX

Boldface indicates illustrations. **I-** indicates color insert.

A

Acheulean tools 102
Africa
   genetic diversity 41–42
   map of key sites 16
   migrations to Eurasia 39–40
   origin of modern humans 41–42, 45–46
   see also specific countries and sites
Almost Human (Berger and Hawks) 199, 218
Archaic humans **33, 38**; *see also* Neanderthals
Ardipithecus kadabba 36
Ardipithecus ramidus **32,** 35–36
Australopithecines 21, 101
Australopithecus 35, 37, 40
Australopithecus afarensis
   age estimates 30, 32
   brain size **38**
   family tree **32,** 34
   fossil footprints 30
   Homo naledi comparison **I-6**
   Lucy 30, **31**
Australopithecus africanus
   age estimates 29, 32, 77
   bipedal locomotion 29, 30
   brain size 29, **38**
   in evolutionary "march of progress" 26
   family tree **32**
   fossil discoveries 28–30, 77
   Taung Child 28, **29**
   teeth 29, 30
   tool use 39
Australopithecus anamensis 30, **32**
Australopithecus boisei 101–102
Australopithecus deyiremeda 30
Australopithecus garhi 30
Australopithecus robustus 39
Australopithecus sediba
   age estimates 33, 38
   brain size **38**
   diet 38
   discovery 37–38, 46–48, 199, I-1
   family tree **33**
   hand 39, 56
   Homo comparisons 38
   paleoproteomics 208
   skull **33, 47**
   teeth 38, 56, 208
   tool use 39

B

Baboons 35
"Backyard syndrome" 155–156
Berger, Jackie 95, 116, 120, 129, 146, 181
Berger, Lee
   Almost Human (with Hawks) 199, 218
   Chute ascent 169, 171–177, **176, I-9, I-16**
   Chute descent 9–12, 126–131, 133–137, 139–146, 219
   Chute descent, preparation for 115–117, 123–124
   Dinaledi Chamber expedition 82–83, 115–117, 159–163, 181, 184, 219, **I-13, I-16**
   Dragon's Back Chamber expedition 117–121, 123–124, **I-4**
   Gladysvale Cave excavations 46
   Hill Antechamber 147–150, 159–160, 219
   Lesedi Chamber expedition 71–76
   markings, discovery of 150–157, **152, 153,** 163–168, **164, 166, 167,** 181
   open-access policy 201
   Rising Star Chamber expedition 192–195
Berger, Matthew 37–38, 46–47, 117, 129
Berger, Megan 117, 129
Berger Box, Rising Star cave system 76, 115
Blombos Cave, South Africa **108,** 156
Bone tools 39
Bonobos 28, 34–35
Boshoff, Pedro 48–49
Brain size
   Australopithecus africanus 29, **38**
   complex behavior and 58
   hominin species **38**

229

Homo naledi **38,** 55, **55,** 56, 58, 80
  mortuary practices and 80, 92, 203
  toolmaking and 101–102
Breccia 18–19, 118
Burials
  contentiousness about *Homo naledi* 103–104, 109–110, 203
  decay and collapse 104–106
  Dinaledi Chamber evidence 89–99, 104–107, **106,** 219, **I-12**
  earliest known 92
  grave goods 98–99, 102–103, 109, 201
  Hill Antechamber evidence 84, **85,** 104, 219
  sophistication necessary for 91–92
  as uniquely human behavior 92, 203

C

Cappellini, Enrico 207–208
Caves
  dead zones 21–23
  living zones 21–22
  prehistoric use of 21, 23
  "touched" zones 21, 23
Charcoal *see* Fire use, evidence of
Chert 18, 63, 65, 107
Chimpanzees 28, 34–35, 101, 204
China
  hominid fossils 41
  Homo erectus 39

Chute, Rising Star cave system
  Berger's ascent 169, 171–177, **176, I-9, I-16**
  Berger's descent 9–12, 126–131, 133–137, 139–146
  Berger's preparations for 115–116, 123–124, 128–129
  in cave system diagram **I-10, I-11**
  Chute Troll 66, 137
  discovery of 48, 142
  entrance 65–66, 137
  name 143
  network of potential passages 142, 184
  presumed tumble of bones down 81–83
  safety and rescue plans 66, 124
  widening 144–145
  width 66, 86, 140, 143
Command Center, Rising Star cave system
  in cave system diagram 20, **I-11**
  Chute descents 66, 145
  communications 49, 66, 67
  Dinaledi Chamber expedition 130
  Dinaledi Puzzle Box expedition 89
  Dragon's Back Chamber expedition **I-4**
  equipment 9, 49, 64, 67
  Lesedi Chamber expedition 67
  location 64
Cradle of Humankind 16–23, **I-1**
  map 16
  see also specific sites

Cro-Magnons 26
CT scans, Hill Antechamber materials **94,** 94–99, **98,** 102, 107

D

Dart, Raymond 28
Dating methods
  electron spin resonance (ESR) 60
  Homo naledi 59–60
  paleoproteomics 207–208
  radiocarbon dating 59–60
  rock carvings 202
  uranium-series dating 60
Denisovans **33,** 41
Dinaledi Chamber, Rising Star cave system
  Berger in 160–163, **I-16**
  Berger's training for 115–117
  bones, deposition theories 81–83, 89–91, 110, 148
  bones, inventory of 117
  burial chamber, evidence of 57–59, 78, 89–99, 104–107, **106,** 219, **I-12**
  in cave system diagram 20, **I-10, I-11**
  discovery of 48–49, 218
  expedition members 217
  expeditions 81, 105, **106,** 120, 130, **I-8**
  fire evidence 194

INDEX

fossil discoveries 10, 66, 77–78, 81, 90, **106**, 120, **I-2**, I-8
Hill Antechamber passage 150–157, **152, 153**, 159–160
mapping 82
markings **182, 183**, 184–186, 201–202, **I-13**
name meaning 56
pathway into 64–65, 118–119, **I-10–I-11**
Puzzle Box 51, 89–90, 104
scientific instruments 150
stalactites and stalagmites 22, 150
time line 218–219
Diversity, evolution of 34–36
Dmanisi, Republic of Georgia 39
DNA evidence
from fossils 208
genetic diversity 41–42
hominid evolution 40–41
primate evolution 27–28
Dolomite
hardness 63, 163
Hill Antechamber stone tool 107
Rising Star cave markings 150–157, **152, 153**, 163–168, **164, 166, 167**
Dragon's Back Chamber, Rising Star cave system
in cave system diagram 20, **I-10, I-11**
Dragon's Back ridge 65, 119, **125**, 136
expeditions 117–121, 123–124, 130, 133–136, 219, **I-4, I-5, I-9**
fire evidence 187–190
fossils 118, 119–121, **I-15**
Homo naledi in 118–119
safety line 65, 134
size 65, 118
time line 219

E

Early Man (book) 26–27
Electron spin resonance (ESR) 60
eLife (journal) 57, 76, 199, 218
Elliott, Marina **I-1, I-5**
ESR (electron spin resonance) 60
ESRF (European Synchrotron Radiation Facility), Grenoble, France 95, 107
Ethiopia: paleoanthropology 16, 30, **31**, 35, 39
Eurasia, *Homo erectus* dispersal into 39
Europe *see* Cro-Magnons; Neanderthals
European Synchrotron Radiation Facility (ESRF), Grenoble, France 95, 107
"Everest Shuffle" 173–174
Evolution *see* Human evolution

F

Fichtel, Stefan I-7
Fire use, evidence of contentiousness 134–135

Dinaledi Chamber 194
Dragon's Back Chamber 134–135, 187–190
first use 188
Homo erectus 135, 188
Homo naledi 67, 134–135, 187–190, 194–198, **194, 196**, 201, 219, 220, **I-14**
Lesedi 67
Rising Star cave system 201, 219
Rising Star Chamber 194–198, **194, 196**, 220, **I-14**
Swartkrans (site), South Africa 189
Flores (island), Indonesia 40; see also *Homo floresiensis*
Flowstone **18**, 18–19, 60–61, 149–150
Footprints, fossil 30
Fuentes, Agustín
background 109, 121
Berger's Chute descent 126, 130, 184
Dragon's Back Chamber expedition 121, 123–124, **127, I-4**
invited to collaborate on *Homo naledi* 109
rock markings and 184–185, 202

G

Genetics *see* DNA evidence
Georgia, Republic of: *Homo erectus* 39
Gladysvale Cave, South Africa 16, 46
Goodall, Jane 101
Gorham's Cave, Gibraltar **182, 183,** 184–185, 202, **I-13**
Gorillas 204

231

Grave goods 98–99, 102–103, 109, 201
Gray, Tom 30
Gurche, John, artwork by **58, I-3, I-6**
Gurtov, Alia 71–76

H

Hawks, John
  Almost Human (with Berger) 199, 218
  Berger's Chute descent and ascent 126, 130, 133, 134, 176, 184
  childhood 25–26
  Dinaledi markings as similar to Gorham's Cave, Gibraltar 184–185
  Dragon's Back Chamber expedition 117–121, 123–124, **128, I-4**
  genetics research 27–28
  Hill Antechamber CT scans 95–99, 107
  on human evolution 25–43
  Rising Star Chamber expedition 192
  Rising Star fire evidence 134
  Skylight Chamber 71
Highveld region, South Africa 17–18
Hill, Lyda 83
Hill Antechamber, Rising Star cave system
  Berger's exploration 147–150
  broken flowstones 149–150
  burial chamber, evidence of 84, **85,** 104, 219
  in cave system diagram 20, **I-11**
  child's teeth **94,** 97, **98**
  doorway to Dinaledi Chamber 150–157, **152, 153**
  excavations 83–87, **85, I-5**
  passage to Dinaledi Chamber 159–160
  skeleton, CT scans **94,** 94–99, **98,** 107
  skeleton recovery 84, 86–87
  stone tools 102–103, 107–109, **108,** 201
  time line 219
Hominins *see* Human evolution; *specific species*
Homo
  earliest fossils found 39
  evolution 37–39
  interbreeding 41
  Sterkfontein (site), South Africa 77
  toolmaking 39
Homo antecessor **33,** **38,** 40, 41
Homo erectus
  Acheulean tools 102
  age estimates 40
  body shape 39–40
  brain size **38**
  caves and 21
  cooperative hunting and gathering 40
  dispersal into Eurasia 39
  etchings by 156
  evolution 26, 39–40, 42
  family tree **33**
  fire, use of 135, 188
  Homo naledi comparison **I-6**
  interbreeding 41
  skull **33**
Homo floresiensis **33, 38,** 40
Homo habilis **33, 38,** 39, 40, 101–102
Homo luzonensis 40
Homo naledi
  age estimates 42–43, 59–61, 93, 149–150, 199
  Australopithecus and *Homo* comparison **I-6, I-7**
  behavior 62
  bipedal locomotion 54, 103, **I-6**
  brain size **38,** 55, **55,** 56, 58, 80
  burials, evidence of 103–107, 109–110, 201, 203, 204
  as cave dwellers 77–80
  classification 54–56, 57
  climbing ability 55, 56, 189
  coexistence with *Homo sapiens* 199, 203, 204–205
  culture, evidence of 198, 201–205
  dating of 59–61, 93
  diet 189, 195
  digital sharing of fossils 201
  discovery (2013) 48–51, 54
  Dragon's Back Chamber 118–121
  face 51
  family tree **33**
  feet 51, **52,** 54, 96, **I-6, I-7**
  femur 51, 66–67, 90, 96
  fingers 51, 54–55, **I-7**
  fire, use of 67, 134–135, 187–190, 194–198, **194, 196,** 201, 219, 220, **I-14**
  hands 50, 51, **53,** 54–56, 84, **I-6, I-7**
  jawbones 49–50, **50,** 51, 74, 84, 90

INDEX

legs 51, 54, 96, **I-7**
markings made by
   150–157, **152, 153,**
   163–168, **164, 166,**
   167, 182, 183, 184–
   186, **197,** 201–202,
   220
mortuary practices
   57–59, **61,** 78, 80,
   91–92, 204
name meaning 10, 56
number of bones and
   teeth recovered 200
paleoproteomics 208
physical portrait 200
portrait of **58**
public announcement
   I-8
scientific reaction to
   discovery 57–59
size 54, 163
skeleton (child's) **94,**
   96–99
skeleton (composite)
   I-7
skeleton (Dinaledi
   Chamber) **I-2**
skeleton (Hill
   Antechamber)
   84–87, **85**
skeleton (Neo) 77, 78,
   79, 80, 94, 218, 219
skull **33,** 51, 55, **55,**
   73, **73,** 74, 76, 84,
   I-7, I-8
skull reproduction **I-8,**
   I-14
specific uses for cave
   chambers 198, 200
stone hearth 195–198,
   196, 197, 201, 220
stone tools 98–99,
   102–103, 107–109,
   108, 201
teeth 54, 55–56, 57, 60,
   74, 83–84, **94,** 95,
   97, **98,** 208

thighbones 50
time line of discoveries
   218–220
wrists 51, 84
Homo rudolfensis **33, 38,**
   39
Homo sapiens
   brain size **38**
   burials 92
   coexistence with *Homo*
     naledi 199, 203,
     204–205
   evolution 40, 61
   in evolutionary "march
     of progress" 26–27,
     26–27, 203–206
   family tree **33**
Human burial *see* Burials
Human evolution 25–43
   Africa as birthplace
     45–46
   diversity 34–36
   DNA evidence 40–41
   Homo (genus) 37–40
   "march of progress" **26–**
     27, 188, 203–205
Human migrations 39–40
Hunter, Rick
   Chute, discovery of 142
   Dinaledi Chamber
     expeditions 48, 81,
     120, 218
   Lesedi Chamber
     expedition 66–67,
     71–76, 120

I
Indonesia: paleoanthro-
   pology **33, 38,** 39, 40
Iron Age 196
Israel: early human burials
   92

J
Johanson, Donald 30
Johnson, Sarah **I-15**

K
Kenya
   fossil discoveries 30, 36
   human burial 92
   key sites map 16
   stone tools 38–39
Kenyanthropus 39
Kenyanthropus platyops 30,
   32, 34
Kruger, Ashley **I-5**

L
Laetoli, Tanzania 16, 30
Late Stone Age 196
Leakey, Mary 30
Leakey, Meave 30
Ledi-Geraru, Ethiopia 16,
   39
Lesedi Chamber, Rising
   Star cave system
   as burial site 93–94
   in cave system diagram
     20
   discovery of 218
   expeditions 67, 71–76
   fire evidence 67
   fossils **73,** 73–74,
     76–78, 120
   name meaning 67
   route to 71–72
   time line 218–219
Letimela (skull) 219, **I-14**
Lomekwi, Kenya 16, 38–
   39
Lucy (*Australopithecus*
   skeleton) 30, **31**
Luzon (island), Philippines
   40

M
Madupe, Palesa 207–208
Makapansgat, South Africa
   16, 28–29
Malapa (site), South Africa
   16, 46; see also
   Australopithecus sediba

Maps
    Cradle of Humankind 16
    Dinaledi Chamber 82
    mapping technology 82
    Rising Star cave system 20, 81–82, **I-10–I-11**
MicroCT scanner 94
Migrations, human 39–40
Modern humans see *Homo sapiens*
Molopyane, Keneiloe "Bones"
    Berger's Chute ascent 176
    Berger's Chute descent 126
    Dinaledi Chamber, inventory of bones 117
    Dinaledi Chamber expedition (2022) 105, 107
    Dinaledi Puzzle Box excavations 89–91, 92
    Dragon's Back Chamber expedition 119, 121, 130, 133, **I-9**
    Dragon's Back Chamber fire evidence 187–190
    expertise 89
    Rising Star fire evidence 134–135
Morris, Hannah 140
Mortuary practices
    defined 78
    grave goods 98–99, 102–103, 109, 201
    Homo naledi 57–59, **61**, 78, 80, 91–92, 204
    oldest instances 78, 80
    as uniquely human 78, 80, 92, 203
    see also Burials

N
National Geographic Society Explorers Festivals 172–173
Neanderthals
    age estimates 33, 40
    brain size **38**
    burials 92
    evolution 26, 40–41, 42
    family tree **33**
    interbreeding 41
    markings made by **182, 183,** 184–185, 202, **I-13**
    range 40
    skull **33**
Neo (*Homo naledi* skeleton) 77, 78, **79,** 80, 94, 218, 219
Noble, Erica **I-5**

O
Oldowan tools 101–102
Open-access channels 201
Orrorin tugenensis 36

P
Paleoproteomics 207–208
Panga ya Saidi cave, Kenya 92
Paranthropus **38,** 39
Paranthropus aethiopicus **32**
Paranthropus boisei **33,** 37
Paranthropus robustus **33,** 37, 50
"Pathfinder syndrome" 142–143
Philippines: paleoanthropology 40
Photogrammetry 74

R
Radiocarbon dating 59–60
Ramalepa, Maropeng
    Berger's Chute ascent 176–177
    Berger's Chute descent 11, 126, 136, 137, 139–140, 144
    Dinaledi Chamber expedition 134
Ramaphosa, Cyril **I-8**
Ramsawak, Ginika **I-5**
Reinhard, Johan 173
Rising Star cave system, South Africa
    Berger Box 76, 115
    burials, evidence of 57–59, **61,** 78, 80, 89–99, 103–107, **106,** 109–110, 201, 204, 219, **I-12**
    climbing difficulties 11–12, 71–76, 86, 87
    connected caverns 63–64
    dating 60–61, 63, 93
    expedition (2013) 48–49, 66–67, 81
    expedition (2017–2018) 81–87
    expedition (2018) 89–91, 92–93
    Exploration Center 126, 130
    fire evidence 67, 134–135, 187–190, 194–198, **194, 196,** 201, 219, **I-14**
    flowstone 60–61, 149–150
    Landing Zone 81, 82, 83
    on map 16
    maps of 20, 81–82, **I-10–I-11**
    markings 150–157, **152, 153,** 163–168, **164, 166, 167, 182, 183,** 184–186, **197,** 201–202, 220, **I-13**
    number of bones recovered from 200

234

INDEX

Rising Star Chamber 191–198, **194, 197,** 220, **I-14**
size 19
Skylight Chamber 64, 67, 71, 89, **I-1**
stalactites and stalagmites **18,** 19
stone hearth 195–198, **196, 197,** 201, 220, **I-14**
stone tools 98–99, 102–103, 107–109, **108,** 201
Superman's Crawl 65, 115, 133, **I-10**
Toilet Bowl 72
Upside-Down Turnaround 20, 192
see also Chute; Command Center; Dinaledi Chamber; Dragon's Back Chamber; Hill Antechamber; Lesedi Chamber
Russia: Denisovans 41

S

Sahelanthropus tchadensis 36
Sesotho (language) 10, 56, 67, 77, 79
Skylight Chamber, Rising Star cave system 64, 67, 71, 89, **I-1**; *see also* Command Center
Smart, Warren
  Berger's Chute ascent 176
  Chute descent with Berger 126–127, 130, 134, 135–137
  Hill Antechamber 149
Spikins, Penny 109–110, 155

Stalactites 18, **18,** 19, 150, 192–193
Stalagmites 18, 19
Sterkfontein (site), South Africa 16, 28–29, 77
Stone tools
  Blombos Cave, South Africa **108**
  earliest 38–39
  Homo naledi 98–99, 102–103, 107–109, **108,** 201
  3D-printed replica 102
Stromatolites 164–165, **I-13**
Swartkrans (site), South Africa 16, 189
Synchrotron scans 107

T

Tanzania: paleoanthropology 16, 30
Taung, South Africa 16
Taung Child (*Australopithecus africanus* skull) 28, **29**
Taurozzi, Alberto 207–208
3D-printed tool replica 102
3D reconstruction of Dinaledi Chamber burial feature **I-12**
3D reconstructions of caves 74
3D scanning of fossils 94–95
3D virtual model of route to Dinaledi 82
Time line of *Homo naledi* discoveries 218–220
Toolmaking 39, 101–102; *see also* Bone tools; Stone tools
Tsikoane, Mathabela **18,** 137, 168, **I-4**
Tucker, Steve

Chute, discovery of 142
Dinaledi Chamber expeditions 48–49, 81, 82–83, 120, 218
Lesedi Chamber expedition 66–67, 71–76, 120
Turkana, Lake, Kenya 16, 30, 38–39

U

University of the Witwatersrand, Johannesburg, South Africa 28
Uranium-series dating 60

V

Van Rooyen, Dirk
  Berger's Chute ascent 172
  Berger's Chute descent 126, 134, 136–137, 140–141, 143–145
  Dinaledi Chamber expedition 130, 134
  Dragon's Back Chamber expedition **I-4**
  Hill Antechamber 149, **I-5**
  Rising Star Chamber expedition 192–195
  Rising Star markings 154–155
Victoria, Lake, Kenya 16, 39

W

Witwatersrand region, South Africa 17–18

235

ABOUT THE AUTHORS

Lee R. Berger, Ph.D., D.Sc., a National Geographic Explorer in Residence, is an award-winning researcher, explorer, author, and speaker. He is the recipient of the National Geographic Society's first Prize for Research and Exploration and the Academy of Achievement's Golden Plate Award. He is an honorary professor and director of the Centre for the Exploration of the Deep Human Journey at the University of the Witwatersrand, and a Senior Carnegie Science Fellow. He continues to lead excavations at both the Malapa and Rising Star sites outside Johannesburg, South Africa.

Berger's work at Rising Star has resulted in the discovery of the largest primitive hominin fossil assemblage in history. His explorations into human origins on the African continent, Asia, and Micronesia for the past two and a half decades have resulted in many discoveries, including the discovery of two new species of early human relatives: *Australopithecus sediba* and *Homo naledi*. Berger is a recognized proponent of open-access science and open sourcing. His novel approach to inclusive science and open collaboration has given him recognition as a Pioneer in Science by the World Science Festival and, in 2016, recognition as one of the 100 Most Influential People in the World from *Time* magazine. He lives in South Africa with his family.

John Hawks, Ph.D., is an internationally recognized expert on human evolution and genetics. He has been a faculty member at the University of Wisconsin–Madison since 2002, where he is presently the Vilas-Borghesi Distinguished Achievement Professor of Anthropology. He is also a visiting professor at the University of the Witwatersrand. In addition to his work on *Homo naledi* at Rising Star and other fossil sites in South Africa, he is known for his work on the genetics of humans' ancient relatives. His work has shown the rapid evolution of modern humans in the last phase of our evolution and has emphasized the genetic connections between Neanderthals and recent people. He has done fieldwork in Africa, Asia, and Europe, combining skeletal evidence from fossils with new information from genetics to uncover how humans evolved. He lives outside Madison, Wisconsin, with his wife, Gretchen, and their four children.